Sensible Decision-Making

Yoshiteru Nakamori

Sensible Decision-Making

Cultivating Systemic and Creative Thinking

Springer

Yoshiteru Nakamori
School of Knowledge Science
Japan Advanced Institute of Science
and Technology
Nomi-shi, Ishikawa, Japan

ISBN 978-981-96-6810-6 ISBN 978-981-96-6811-3 (eBook)
https://doi.org/10.1007/978-981-96-6811-3

This Springer imprint is published by the registered company Springer Nature Singapore Pte Ltd.
The registered company address is: 152 Beach Road, #21-01/04 Gateway East, Singapore 189721, Singapore

Preface

We often prioritize rationality in decision-making, valuing logic and evidence as essential tools for guiding our choices. While rationality is indispensable, it alone may not be sufficient in complex and uncertain situations. Sensibility—the ability to intuitively grasp nuances and adapt to changing circumstances—also plays a crucial role in making truly effective decisions. Thus, this book emphasizes *Sensible Decision-Making*, a concept that integrates both rational analysis and intuitive wisdom. To make well-rounded decisions, we must strive not only for sound decisions that are logically correct and consistent but also for sensible decisions that are adaptable and intuitively appropriate. Achieving this balance requires more than just rational thinking; it also necessitates *Cultivating Systemic and Creative Thinking*.

In the mid-twentieth century, mathematical methods such as operations research and systems engineering emerged, giving rise to *rational decision-making*—a process centered on logical reasoning and objective data. This approach, rooted in ancient Greek philosophy and later expanded during the scientific revolution and within economic theory, is explored in Chapter 1. This chapter examines decision-making under uncertainty, game theory for strategic decision-making, and multi-objective decision-making. Rational decision-making offers numerous benefits, such as its reliance on objective data, the clarification of decision criteria, and the systematic selection of the best option. However, it also has limitations: incomplete information, ethical dilemmas, and cultural differences can hinder its real-world applicability.

As awareness has grown that decision-making cannot be entirely detached from subjective factors, more holistic approaches to decision-making have emerged across various disciplines, particularly systems science and management science, integrating emotions and perceptions with rationality. Chapter 2 introduces Affective (Kansei) Engineering and Emotional Design, which harness human sensibility to create impactful products and services. This book uses the term *sensibility* to capture the essence of the Japanese concept of *Kansei*. To effectively incorporate sensibility into decision-making, a rational framework is essential. Mathematical methods, such as fuzzy set theory, have been utilized to model subjective concepts. Chapter 2 proposes simple *possibility models*, derived from fuzzy set theory, to represent human sensibility and presents practical decision-making methods that aggregate quantified aspects of sensibility using these models.

One of the key challenges in rational decision-making is accurately modeling all relevant factors. While holistic problem-solving is ideal, systemic thinking provides a more practical alternative by emphasizing interconnected relationships and feedback loops. *Systemic decision-making* offers a broader and more dynamic perspective than traditional methods. Chapter 3 introduces tools such as causal loop diagrams and soft systems methodologies, which facilitate systemic decision-making by addressing the complex interplay of factors and feedback mechanisms. These approaches promote a comprehensive understanding of problems. However, they also have limitations, such as difficulties in modeling nonlinearities, managing uncertainties, and handling constraints related to data availability, time, and human cognitive capacity.

Despite these challenges, humans possess the remarkable ability of creative imagination. *Creative decision-making* is expected to drive innovation by generating original and valuable outcomes through the integration of intuition and logical reasoning. Chapter 4 introduces knowledge creation models and knowledge management, which emerged in the final decade of the twentieth century and have since become widely adopted, driven by growing expectations for a knowledge-based society. These approaches leverage human experiential knowledge to complement data-driven, objective methods. This chapter also examines structured techniques for fostering creativity, such as mind mapping and design thinking, which have been formalized into widely adopted business frameworks.

While creative decision-making is essential for generating innovative ideas, it comes with inherent challenges. Objectively assessing the value and feasibility of new ideas can be complex. Creative ideas often challenge the status quo, increasing the risk of failure, and their implementation may face resistance to change, resource constraints, or technological barriers. Rational, systemic, and creative approaches each have distinct strengths but also inherent limitations. This book introduces synthetic decision-making as a way to overcome these challenges and harness the strengths of all three approaches.

Though still an emerging field, *synthetic decision-making* aims to create new value and enhance decision-making by integrating diverse methodologies. Chapter 5 explores synthetic thinking, expanding on the systemic and creative thinking frameworks introduced in Chaps. 3 and 4. It distinguishes between thought processes driven by rational knowledge and those guided by intuitive insights. This chapter presents a novel decision-making approach that synthesizes these two forms of knowledge, aligning with the book's central theme of *Sensible Decision-Making*. Chapter 5 also introduces *the Knowledge Triad Synthesis Framework*, which supports this integration, drawing from the author's research on knowledge construction methodology. The knowledge triad highlights the fusion of rational and intuitive knowledge facilitated by the connective power of social knowledge.

This book examines rational, systemic, and creative approaches and ultimately advocates for integrating these methods through synthetic thinking. While Chapter 5 presents the author's perspective, the book objectively analyzes contemporary decision-making methodologies. As technology and society continue to evolve, decision-making is undergoing profound transformations. Emerging approaches may incorporate human-AI collaboration, data-driven simulations, virtual reality, collective intelligence, and insights from neuroscience. As you engage with the concepts in this book, consider how AI and other technological advancements might shape the future of decision-making.

Ishikawa, Japan Yoshiteru Nakamori

Competing Interests The author has no competing interests to declare that are relevant to the content of this manuscript.

Contents

Chapter 1
Decision-Making with the Logical Mind

Abstract This chapter explores the foundational principles of rational decision-making, in which logic and evidence play central roles. Rational decision-making involves selecting actions that maximize the likelihood of achieving specific goals. However, its effectiveness depends on the quality and completeness of the information used to define and analyze the problem. This chapter reviews traditional frameworks, including decision-making under uncertainty, game theory for strategic decision-making, and multi-objective decision-making. It critically examines their underlying assumptions and explores their strengths and limitations in real-world contexts.

1.1 Logical Thinking

The spread of the Internet has led to an abundance of information, which is often conflicting and overwhelming. Opportunities to interact with people from different cultures and with diverse values are increasing. Flexible and logical thinking is crucial for adapting to these social changes and developing new strategies. Logical thinking aims to make optimal decisions by basing judgments on objective data while minimizing the influence of biases and emotions.

There are several reasons why logical thinking is essential:

- It is a powerful tool for structuring complex problems, analyzing them logically, and arriving at optimal solutions.
- It enables you to analyze data objectively, make rational decisions based on the results, and adapt effectively to changing circumstances.
- It helps you communicate your ideas clearly and logically, making it easier to persuade others.
- It facilitates constructive discussions with team members, leading to the generation of better ideas.

Y. Nakamori, *Sensible Decision-Making*, https://doi.org/10.1007/978-981-96-6811-3_1

1.1.1 Logic and Rationality

Logical thinking primarily concerns reasoning and inference. It involves evaluating evidence, identifying patterns, and drawing conclusions based on established rules and principles.

Rational decision-making, on the other hand, extends beyond mere logic. It involves:

- Clarifying goals and objectives: Defining what you want to achieve.
- Evaluating options: Assessing the potential outcomes of different choices in light of your goals.
- Selecting the best course of action: Selecting the option most likely to help achieve your desired outcome.

In summary, while logic provides the foundation for understanding and analyzing information, rational decision-making involves applying that understanding to make purposeful, effective choices that align with your goals. *Logical reasoning serves as the basis for rational thinking.* However, *being logical does not necessarily mean being rational.*

1.1.1.1 Logical Thinking Tools

Logical thinking is a powerful cognitive process that helps us make sense of the world around us. It involves using *reason and evidence* to analyze information, identify relationships, and draw sound conclusions. Here are two logical thinking tools commonly used in business settings:

MECE Analysis[1]: MECE stands for "*Mutually Exclusive and Collectively Exhaustive*," meaning "without omissions or duplications." MECE analysis breaks down problems and issues into mutually exclusive and collectively exhaustive components and organizes them systematically.

MECE analysis helps ensure that your thinking is clear, comprehensive, and structured. Breaking down complex problems into well-defined categories allows you to:

- Identify and address all potential causes of an issue.
- Avoid overlooking critical factors.
- Ensure that all aspects of a situation are considered.
- Make better, more focused decisions.

Examples of MECE analysis in action include:

- Analyzing customer segments: Based on purchase history and spending patterns, customers can be categorized as *high-value*, *mid-value*, or *low-value*.

---

[1] Barbara Minto developed the MECE principle in the late 1960s at McKinsey & Company.

- Identifying marketing channels: Marketing channels can be classified as *online* (social media, email, search engine marketing) and *offline* (print advertising, television commercials, direct mail).
- Diagnosing a sales slump: Possible categories include factors related to the product, the sales team, or market conditions.

SWOT Analysis[2]: SWOT analysis is a framework for evaluating a company's Strengths, Weaknesses, Opportunities, and Threats. It helps you assess both your internal capabilities and external environment.

- *Strengths*: Internal, positive attributes that provide a competitive advantage. These may include skills, experience, knowledge, resources, or a strong reputation.
- *Weaknesses*: Internal factors that may hinder performance. These could involve a lack of skills, resources, or experience, as well as unfavorable aspects of your situation.
- *Opportunities*: External factors that present favorable circumstances you can leverage, such as emerging market trends, unmet customer needs, or competitors' weaknesses.
- *Threats*: External challenges that may pose risks, including economic downturns, regulatory changes, increased competition, and technological advancements.

Applications of SWOT analysis include:

- Business planning: Companies use SWOT analysis to evaluate their competitive position and develop strategies for growth.
- Product development: Teams apply SWOT analysis to assess new product ideas, identify strengths and weaknesses, and explore market opportunities and threats.
- Personal development: Students use SWOT analysis to evaluate their academic strengths and weaknesses, improve study habits, and consider potential career paths.

1.1.1.2 Induction and Deduction

Logical thinking is a powerful cognitive process that helps us understand the world around us. It involves using reason and evidence to analyze information, identify relationships, and draw sound conclusions. There are two fundamental methods of logical reasoning:

- *Induction*: This method draws conclusions based on multiple observations or instances. It is a form of probabilistic reasoning—while numerous cases may support a conclusion, exceptions are always possible. Inductive reasoning is useful for identifying patterns and generalizing, thereby generating new knowledge.
- *Deduction*: This method derives logically necessary conclusions from given premises, assuming those premises are true. It follows the principles of deductive logic and does not rely on empirical evidence. Unlike induction, deduction

[2] The Stanford Research Institute developed the SWOT analysis during the 1960s and 1970s.

Table 1.1 Comparison of inductive and deductive reasoning

Feature	Inductive reasoning	Deductive reasoning
Direction of reasoning	Specific to general	General to specific
Certainty of reasoning	Uncertain/probable	Certain (if premises are true)
Strength of reasoning	Weak (may lead to false conclusions)	Strong (conclusions logically follow from premises)
Purpose	Discover patterns, form hypotheses, and generate new knowledge	Test hypotheses, apply rules, and solve problems
Real-world application	Scientific research, market research, predicting future trends	Mathematical proofs, legal reasoning, diagnosis
Key proponent	Francis Bacon	Aristotle (commonly associated with deductive logic)

does not produce new knowledge; rather, it applies existing principles to reach a specific, logically sound conclusion.

Table 1.1 presents a clear comparison of inductive and deductive reasoning.

Below are some simple examples of inductive and deductive reasoning:

1. *Inductive Reasoning* (Scientific Observations):
    - Observation: Scientists observe that every swan they have encountered is white.
    - Conclusion: All swans are white.
    - Note: This famous example illustrates that inductive reasoning can lead to incorrect conclusions if the sample size is limited.
2. *Deductive Reasoning* (Medical Diagnosis):
    - Proposition: If a patient has the flu, they will typically exhibit fever, chills, and body aches.
    - Observation: This patient exhibits fever, chills, and body aches.
    - Conclusion: This patient may have the flu.
    - Note: If the proposition is not a universal law, the conclusion will not be 100% certain. Medical diagnoses require more comprehensive analysis.

1.1.1.3 Demand Forecasting: A Blend of Induction and Deduction

Demand forecasting is a critical business activity that estimates future demand based on historical data. This process effectively incorporates both inductive and deductive reasoning, each offering distinct advantages and challenges. Table 1.2 summarizes the roles, benefits, and challenges of these reasoning approaches in demand forecasting.

Table 1.2 Benefits and challenges of inductive and deductive reasoning in demand forecasting

Feature	Inductive reasoning	Deductive reasoning
Roles	• *Analyzing historical sales data*: Identifying patterns, trends, and seasonality in past sales figures • *Market research and observation*: Gathering data on customer behavior, competitor activities, and market trends • *Developing forecasting models*: Using statistical methods like time series analysis, regression analysis, and moving averages to predict future demand based on historical data	• *Using economic theories and models*: Applying established economic principles (e.g., consumer theory, supply and demand) to predict demand • *Scenario planning*: Developing different scenarios based on assumptions about economic conditions, competitor actions, and other relevant factors • *Expert judgment and opinion*: Incorporating insights from industry experts, market analysts, and other knowledgeable sources
Benefits	• *Data-driven approach*: Relies on empirical evidence and historical data • Identifies emerging trends and patterns • Forecasts demand for a wide range of products and services	• Provides a framework for understanding underlying market forces • Explores different scenarios and their potential impacts • Incorporates expert knowledge and insights
Challenges	• Accuracy depends on the quality and availability of historical data • May not accurately predict unexpected events or sudden changes in market conditions • Susceptible to biases in data collection and analysis	• Relies on the accuracy and validity of underlying economic theories and assumptions • May not be effective in rapidly changing or unpredictable markets • Heavily influenced by subjective expert opinions

Table 1.2 suggests that both inductive and deductive reasoning incorporate quantitative and qualitative approaches in demand forecasting:

1. Inductive Reasoning

   - Quantitative approach: Develops forecasting models from numerical data using statistical methods (e.g., analyzing historical sales data to identify trends and patterns).
   - Qualitative approach: Identifies market trends through surveys on consumer needs and competitor analysis (e.g., conducting market research to understand consumer preferences).

2. Deductive Reasoning

   - Quantitative approach: Generates numerical forecasts through scenario analysis based on established models (e.g., applying a regression model to predict sales under different economic conditions).

- Qualitative approach: Develops demand forecasts based on market trends, drawing on economic theory and expert opinions (e.g., forecasting demand based on anticipated changes in consumer behavior and expert assessments).

Both quantitative and qualitative approaches are essential for comprehensive demand forecasting. They reflect the human decision-making process, which combines logical analysis with intuitive thinking. However, effectively integrating these approaches poses challenges. Factors such as data accuracy, decision-maker mindsets, and company culture play a crucial role in the success of this integration.

1.1.1.4 A Framework for Knowledge Creation

The ultimate goal is to make better decisions by effectively integrating rational and intuitive insights. This principle extends beyond demand forecasting and applies to all decision-making processes. Success in knowledge management depends heavily on seamlessly integrating codified (structured) knowledge and personalized (tacit) knowledge.

Figure 1.1 illustrates a value-creation spiral model, which represents the process of generating valuable ideas by transforming information into knowledge. The model comprises four key processes, each with a distinct role:

1. *Observation*: Collect data (codified knowledge) and existing knowledge (personalized knowledge) relevant to solving the problem, and then organize them into structured information.
2. *Understanding*: Analyze and convert the information into knowledge, including mathematical predictive models (codified knowledge) and the analyst's contextual understanding (personalized knowledge).

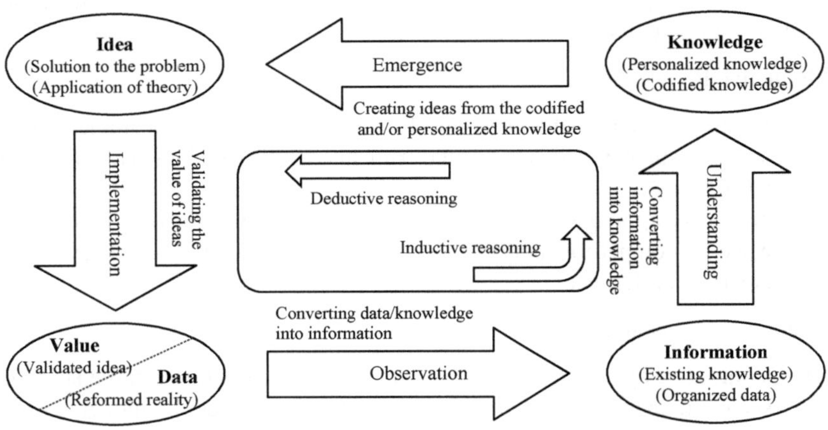

Fig. 1.1 OUEI model—a value-creation framework (Nakamori, 2021)

3. *Emergence*: Generate problem-solving ideas by leveraging the acquired knowledge. These ideas stem from scenario-based predictions (codified knowledge) derived from predictive models and creative inspiration (personalized knowledge) shaped by analysts' expertise.
4. *Implementation*: Apply and validate the ideas. Validation can be performed using quantitative evaluation methods (codified knowledge) and expert assessments (personalized knowledge) conducted by evaluators.

This iterative process, known as the OUEI model—a cyclical framework consisting of *Observation*, *Understanding*, *Emergence*, and *Implementation*—exemplifies a common approach to knowledge creation and decision-making.

When applied to demand forecasting, this model can be interpreted as follows:

- Observation and Understanding (Inductive Reasoning): Focuses on gathering and analyzing structured and unstructured data to build a foundational understanding and develop a demand forecasting model capable of predicting future outcomes.
- Emergence (Deductive Reasoning): Applies acquired knowledge to generate specific solutions. This includes applying the prediction model to hypothetical scenarios to deduce expected outcomes and leveraging analysts' insights to develop adaptive solutions.
- Implementation (Evaluation): Implements ideas to assess both inductive and deductive reasoning. Models can be evaluated using well-established performance metrics, but this process is crucial for gaining key insights that inform the next cycle.

1.1.2 Rational Decision-Making Situations

Rational decision-making involves making choices based on logical reasoning and objective analysis. The primary goal is to choose the option that maximizes the likelihood of achieving desired outcomes while minimizing the influence of emotions and personal biases.

Key steps in the rational decision-making process include:

1. Define Goals: Clearly articulate the specific objectives to be achieved.
2. Identify Options: Generate a comprehensive list of potential courses of action.
3. Gather Information: Collect relevant data and insights about each option.
4. Evaluate Options: Analyze the potential advantages and disadvantages of each option.
5. Compare and Select: Weigh the options against the defined goals and choose the most promising course of action.
6. Implement: Execute the selected option effectively.
7. Evaluate and Adjust: Assess the outcomes and make necessary adjustments or refinements.

Types of decision-making situations include:

A. Decision-Making Under Uncertainty: Occurs when outcomes are unpredictable due to unknown or uncontrollable factors.
B. Strategic Decision-Making: Involves multiple decision-makers whose choices interact and influence the overall outcome.
C. Multi-Objective Decision-Making: Requires balancing multiple, potentially conflicting objectives simultaneously.
D. Group Decision-Making: Entails collective decision-making by individuals, often requiring consensus or majority rule.

An overview is provided below, and the following sections will discuss A, B, and C in detail.

1.1.2.1 Overview of Decision-Making Under Uncertainty

Decision-making under uncertainty involves selecting a course of action when future outcomes are unknown or unpredictable. This is a pervasive challenge in various aspects of life, as most real-world situations contain some degree of uncertainty.

Key characteristics of decision-making under uncertainty are:

- Incomplete Information: Decision-makers often lack complete knowledge of the situation, the potential consequences of their actions, and the probabilities associated with different outcomes.
- Multiple Options: Typically, several actions are available, each carrying potential rewards and risks.
- Subjective Evaluation: Assessing potential outcomes requires subjective judgments based on individual values, preferences, and risk tolerance.

Examples of decision-making frameworks under uncertainty include:

- Expected Value Maximization: Selects the option with the highest expected payoff, considering the potential rewards and their associated probabilities.
- Minimax Regret: Minimizes the potential for regret in the worst-case scenario by choosing the option that would result in the least regret if the least favorable outcome were to occur.
- Satisficing: Involves selecting the first option that meets a predetermined set of acceptable criteria rather than seeking the absolute "best" one.

The most suitable decision-making framework depends on the specific context, the individual's risk tolerance, and the relative importance of various potential outcomes.

1.1.2.2 Essentials of Strategic Decision-Making

Strategic decision-making occurs when the outcome of a decision is influenced not only by the choices of an individual decision-maker but also by the decisions made by others. While it can be considered a subset of decision-making under uncertainty, strategic decision-making specifically focuses on the uncertainty arising from the actions of other decision-makers.

The difference between strategic decision-making and decision-making under uncertainty can be seen in the following example:

- Decision-Making Under Uncertainty: A company makes decisions while anticipating the potential impacts of external uncertainties, such as an economic downturn (e.g., recession, inflation).
- Strategic Decision-Making: A company makes decisions about product development while considering competitive uncertainties, such as potential responses from rival firms, including new product launches or pricing strategies.

Game Theory: Developed by John von Neumann and Oskar Morgenstern, game theory provides a formal framework for analyzing strategic decision-making. It models decision-making situations as "games" in which players (decision-makers) interact and make choices that influence each other's outcomes.

1.1.2.3 Fundamentals of Multi-objective Decision-Making

Multi-objective decision-making (MODM) addresses the challenge of selecting the best course of action when multiple, often conflicting, objectives must be considered simultaneously.

Key characteristics of MODM are:

- Multiple Objectives: The decision-maker aims to achieve several distinct goals.
- Incommensurability: Objectives may be difficult to compare or measure using a common unit (e.g., profit maximization vs. environmental sustainability).
- Conflicting Objectives: Achieving one objective may hinder the achievement of another (e.g., minimizing cost while maximizing quality).
- Pareto[3] Optimality: Typically, there is no single "best" solution. Instead, a set of Pareto-optimal solutions exists, where no solution can improve one objective without worsening another.

Application examples include:

- Product Development: Balancing performance, cost, environmental impact, and user experience.
- Investment Portfolio Selection: Considering return on investment, risk level, liquidity, and ethical considerations.
- Environmental Policy: Developing policies that balance economic growth, environmental protection, and social equity.

[3] Vilfredo Pareto was an Italian economist, sociologist, and philosopher best known for developing the concept of Pareto Optimality.

The following are standard MODM techniques:

- Weighted Sum Method: Assigns weights to each objective based on its relative importance and selects the option that maximizes the weighted sum of the objective values.
- ε-Constraint Method: Optimizes one objective while treating others as constraints by setting predefined acceptable levels for them.
- Goal Programming: Establishes target values for each objective and aims to minimize deviations from these targets to find the best solution.
- Utility Function Method: Constructs a utility function representing the decision-maker's overall satisfaction with achieving different objectives and selects the option that maximizes this utility.

1.1.2.4 Insights into Group Decision-Making

Group decision-making is a collective process in which individuals collaborate to choose a course of action. It brings together diverse perspectives, fosters shared responsibility, and leverages the group's collective intelligence.

Key characteristics are:

- Diverse Perspectives: Multiple viewpoints and areas of expertise are integrated, potentially leading to more creative and robust solutions.
- Shared Responsibility: The decision becomes a collective responsibility, increasing ownership and commitment.
- Complexity: Group decision-making can be time-consuming and requires effective communication, conflict resolution, and facilitation skills to ensure a productive process.

Representative techniques include:

- Brainstorming: A creative technique that encourages the free flow of ideas without initial judgment or criticism (Osborn, 1953).
- Nominal Group Technique (NGT): Participants independently generate ideas, which are then shared, discussed, and prioritized through a structured voting process (Delbecq & Van de Ven, 1971).
- Delphi Method: An iterative process that gathers expert opinions anonymously through a series of questionnaires (Dalkey, 1969).
- Multi-Voting: Allows participants to express their preferences by casting multiple votes on different options to indicate their priorities (Lackner & Skowron, 2023).
- Consensus Building: Emphasizes open communication, active listening, and finding common ground through negotiation and compromise (Fisher et al., 1991; Susskind & Cruikshank, 1987).

1.2 Decision-Making Under Uncertainty

This section explores key concepts and methods for making decisions in uncertain environments. Decision-making problems under uncertainty are characterized by the following factors:

- *Options*: The available courses of action from which the decision-maker can select.
- *States* (Environmental Factors): External conditions or events that can significantly influence the outcomes of the chosen options. These factors are typically beyond the direct control of the decision-maker.
- *Consequences*: The potential outcomes that may result from each option, given the different possible states of the environment.
- *Payoff*: A numerical value assigned to each consequence, reflecting its desirability or utility to the decision-maker. Higher payoffs generally correspond to more preferred outcomes.

Rational decision-making under uncertainty aims to select the option that maximizes the expected payoff, considering the available information, the probabilities of different states occurring (if known), and the decision-maker's objectives.

Knight's Uncertainty: Knight (1921) distinguished between two fundamental types of uncertainty:

1. *Risk*: A type of uncertainty in which the probability of different outcomes can be assessed, albeit with varying degrees of precision. This assessment can be based on:

    - Objective Probabilities: Derived from historical data, statistical analysis, or empirical observations.
    - Subjective Probabilities: Based on expert opinions, personal beliefs, or intuitive judgments.

2. *Strict Uncertainty*: A type of uncertainty in which the probabilities of different outcomes cannot be reliably estimated due to a lack of historical data, the presence of unique and unpredictable factors, or the inherent unknowability of the future. In such situations, traditional probability theory and statistical methods may have limited applicability.

Knight argued that strict uncertainty profoundly impacts decision-making processes. Under strict uncertainty, subjective judgments, psychological factors (such as risk aversion, ambiguity aversion, and emotional biases), and even intuition play significant roles in shaping choices. Traditional economic models often assume rational choice under risk, where individuals make decisions by maximizing expected utility based on known or estimated probabilities. However, Knight's concept of strict uncertainty challenges this assumption, suggesting that decision-making in such environments may deviate significantly from the predictions of standard economic theory.

Decision-Making Criteria: It is important to note that "rational" decision-making does not inherently guarantee optimal outcomes in every scenario. Instead, it focuses on selecting the option that, based on the available information and a chosen evaluation criterion, appears most likely to achieve the desired objective.

This section explores four standard decision-making criteria:

A. *Expected Value Maximization Criterion*: This fundamental criterion in decision-making under risk involves selecting the option with the highest expected value. The expected value is calculated by multiplying the potential payoff of each outcome by its probability and then summing these products.
B. *Maximin Criterion*: This conservative approach focuses on minimizing potential losses. It involves identifying the worst-case scenario for each option and selecting the option with the highest payoff among those worst-case scenarios.
C. *Maximax Criterion*: This optimistic approach aims to maximize potential gains. It involves identifying the best-case scenario for each option and selecting the option with the highest payoff among those best-case scenarios.
D. *Minimax Regret Criterion*: This approach seeks to minimize the potential for regret, which arises from the realization that a different choice would have yielded a better outcome. It involves selecting the option that results in the least regret if the worst possible outcome occurs.

Criteria B–D are often employed in situations of strict uncertainty, where accurate probabilistic assessments are challenging or impossible due to a lack of historical data, the presence of unique and unpredictable factors, or the inherent unknowability of the future.

Subsequent Sections: Subsect. 1.2.1 will delve deeper into the expected value maximization criterion, while Subsect. 1.2.2 will explore the remaining three criteria in more detail, using illustrative examples to enhance understanding.

A Decision-Making Problem: Choosing a Food Stall for a School Festival: To illustrate decision-making under uncertainty, consider a scenario in which a student group plans to operate a food stall at their school festival. The festival typically features a variety of activities, including stage performances, exhibitions, lectures, sports tournaments, and numerous other student-run food stalls, creating a dynamic and competitive environment.

The group faces the critical decision of choosing a menu that maximizes profitability. They must consider several key factors:

- Market Demand: Assessing the popularity and appeal of various food and drink options among fellow students, considering their preferences and potential dietary restrictions.
- Preparation Feasibility: Evaluating the practicality of preparing and serving different menu items within the constraints of the festival environment, considering factors such as available space, equipment, and time limitations.
- Profitability Analysis: Determining potential costs associated with ingredients, supplies, and labor while setting competitive prices to attract customers and ensure healthy profit margins.

- Sales Forecasting: Estimating potential sales volume based on historical data (if available), market research conducted among students, and anticipated festival attendance.
- Environmental Considerations: Accounting for potential weather conditions (e.g., sunshine, rain, snow) and their impact on consumer demand for food items.

Accurately predicting profits in such a dynamic environment can be challenging. To simplify the analysis and illustrate key decision-making concepts, we will utilize a hypothetical payoff matrix (Table 1.3), which presents estimated profit figures for four potential food items (options: Fried Noodles, French Fries, Soft Serve Ice Cream, and Crepes) under four different weather conditions (states: Sunny, Cloudy, Rainy, and Snowy).

From here on, for simplicity, soft serve ice cream will be referred to as *soft serve.* Each weather condition occurs with a certain probability:

- P_1: Probability of Sunny weather
- P_2: Probability of Cloudy weather
- P_3: Probability of Rainy weather
- P_4: Probability of Snowy weather

The sum of these probabilities must equal 1 ($P_1 + P_2 + P_3 + P_4 = 1$), reflecting all possible states of the environment.

Weather Forecasting & Initial Decision-Making: Accurately predicting weather conditions for the festival date, typically two to three months in advance, presents a significant challenge. While long-term forecasts provide general guidance, pinpointing the exact weather on the festival day remains uncertain.

One approach to estimating the likelihood of different weather scenarios involves analyzing historical weather data. By examining weather patterns on the festival date over the past 30 years, we can determine the historical frequencies of different weather conditions (Sunny, Cloudy, Rainy, and Snowy). These frequencies serve as an initial estimate of the probabilities for each weather condition on the festival day.

Before incorporating weather probabilities, consider an initial decision regarding the most suitable food stall option.

Table 1.3 A hypothetical payoff matrix (profit forecast table)

Option	Sunny (P_1)	Cloudy (P_2)	Rainy (P_3)	Snowy (P_4)
Fried Noodles	$500	$300	$200	$0
French Fries	$400	$300	$150	$150
Soft Serve	$800	$200	$0	$0
Crepes	$700	$200	$100	$0

Note The profit values in the payoff matrix are purely hypothetical and do not reflect real-world scenarios. This simplified model allows us to explore various decision-making criteria and strategies for selecting the most advantageous food stall option, providing valuable insights into the decision-making process under uncertainty

- Objective: Assuming the primary goal is to maximize profit, which food stall option would you select without knowing the likelihood of different weather conditions?
- Analysis: Carefully examine the payoff matrix and evaluate each food option's potential risks and rewards.

Now, introduce the following weather probabilities:

- Sunny: $p_1 = 0.2$
- Cloudy: $p_2 = 0.5$
- Rainy: $p_3 = 0.3$
- Snowy: $p_4 = 0.0$

Question: Given these probabilities, how would your initial decision change?

In the following sections, we will explore how to make informed decisions under uncertainty. First, we will examine decision-making with known probabilities, as demonstrated in this example. Then, we will explore three key criteria for decision-making when probabilities are unknown.

1.2.1 Decision-Making Under Risk

This subsection explores the concept of expected value maximization, a fundamental approach to decision-making under risk. While this method provides a logical framework for selecting the option with the highest expected payoff, it is important to acknowledge that human decision-making often deviates from this purely rational model.

1.2.1.1 Expected Value Maximization

Expected payoff maximization is a decision-making approach that chooses the option with the highest expected payoff.

We first introduce the following symbols:

- $a_m, m = 1, 2, \ldots, M$: Options (e.g., Fried Noodles)
- $s_n, n = 1, 2, \ldots, N$: States (e.g., Sunny)
- $g(a_m, s_n)$: The payoff associated with each option-state pair (a_m, s_n)
- $p(s_n)$: The probability of each state, satisfying:

$$p(s_n) \geq 0, \quad \sum_{n=1}^{N} p(s_n) = 1$$

For the option a_m, the expected payoff $EP(a_m)$ is calculated as:

$$EP(a_m) = \sum_{n=1}^{N} p(s_n)g(a_m, s_n)$$

The optimal choice, denoted as a^*, is the option that maximizes the expected payoff:

$$a^* \in \underset{m}{\operatorname{argmax}} \sum_{n=1}^{N} p(s_n)g(a_m, s_n), \quad m \in \{1, 2, \ldots, M\}$$

Here, $\underset{m}{\operatorname{argmax}}$ selects the option(s) a_m that maximize the expected payoff across all possible states. The term "argmax" stands for "argument of the maximum," which identifies the specific value(s) of m (or a_m) that yield the highest expected payoff. This notation allows for cases where multiple options achieve the maximum payoff, meaning $\underset{m}{\operatorname{argmax}}$ represents a set of optimal choices rather than a single option.

Note: The symbols above are essential for clearly describing decision-making problems, and we will continue to use them when learning other criteria. However, to avoid unnecessary complexity, we often omit them from the calculations in examples.

To ensure clarity, let us establish the correspondence between the symbols and the decision problem of selecting a food stall at a school festival:

- Options: a_1= Fried Noodles, a_2= French Fries, a_3= Soft Serve, a_4= Crepes
- States: s_1= Sunny, s_2= Cloudy, s_3= Rainy, s_4= Snowy
- Payoff: (Example) $g(a_2, s_3) = g(\text{French Fries, Rainy}) = \$150$
- Probability: (Example) $p(s_1) = p(\text{Sunny}) = 0.2$

For example, the expected payoff for a_2 (French Fries) is given by:

$$EP(a_2) = p(s_1)g(a_2, s_1) + p(s_2)g(a_2, s_2) + p(s_3)g(a_2, s_3) + p(s_4)g(a_2, s_4)$$

Using the provided weather probabilities—Sunny (0.2), Cloudy (0.5), Rainy (0.3), and Snowy (0.0)—along with the profit data from Table 1.3, we calculate the expected payoff for each food option as follows:

- Fried Noodles: $(0.2 \times 500) + (0.5 \times 300) + (0.3 \times 200) + (0.0 \times 0) = 310$
- French Fries: $(0.2 \times 400) + (0.5 \times 300) + (0.3 \times 150) + (0.0 \times 150) = 275$
- Soft Serve: $(0.2 \times 800) + (0.5 \times 200) + (0.3 \times 0) + (0.0 \times 0) = 260$
- Crepes: $(0.2 \times 700) + (0.5 \times 200) + (0.3 \times 100) + (0.0 \times 0) = 270$

Based on the expected payoff maximization criterion, which selects the option with the highest expected profit, Fried Noodles is the most favorable choice, which can be mathematically expressed as follows:

Table 1.4 Weather occurrence rates on school festival days

Weather	Occurrence rate (%)	Average temperature	Highest temperature	Lowest temperature
Sunny	63.3	16.0 °C	25.5 °C	5.3 °C
Cloudy	23.3	**Precipitation**	**Sunshine hours**	**Snowfall amount**
Rainy	13.3	2.4 mm	5.3 h	–
Snowy	0.0			

$$a^* \in \underset{m}{\operatorname{argmax}}\, EP(a_m) = \{a_1\} = \{\text{Fried Noodles}\}$$

Here, $\{a_1\}$ is a singleton set containing only a_1. If multiple options yield the maximum value, $\underset{m}{\operatorname{argmax}}$ will be a multi-element set, requiring an additional decision-making method to determine the final choice.

Using Historical Weather Data: We can refine our probability estimates by using historical weather data from Table 1.4, which presents the average weather occurrence rates on school festival days over the past 30 years. This table provides average values for temperature and precipitation. While the weather on the festival day is expected to be refreshing, it might still be a bit chilly.

By using these historical occurrence rates as probabilities, we can recalculate the expected profits for each food option as follows:

- Fried Noodles: $(0.633 \times 500) + (0.233 \times 300) + (0.133 \times 200) + (0.0 \times 0) = 413$
- French Fries: $(0.633 \times 400) + (0.233 \times 300) + (0.133 \times 150) + (0.0 \times 150) \approx 343$
- Soft Serve: $(0.633 \times 800) + (0.233 \times 200) + (0.133 \times 0) + (0.0 \times 0) = 553$
- Crepes: $(0.633 \times 700) + (0.233 \times 200) + (0.133 \times 100) + (0.0 \times 0) = 503$

Based on these revised calculations, Soft Serve now has the highest expected profit, making it the optimal choice under the expected value maximization criterion.

1.2.1.2 Expected Value Paradox

Real-world decision-making is influenced by various psychological factors, leading to behaviors that may seem irrational from a purely mathematical perspective. The following discussion presents two examples illustrating the expected value paradox.

Lottery: Table 1.5 presents an example of prize amounts and winning probabilities for a typical Japanese lottery (a "Jumbo Lottery"). This lottery involves a group number and a six-digit number.

Let us say lottery tickets cost \$3 each. The expected value of a \$3 lottery ticket is calculated by summing the products of the prize amount and its corresponding

Table 1.5 Winning amounts and winning probabilities for a particular lottery

Lottery grade	Winning amount (A) (Unit: US$)	Winning probability (B)	(A) × (B) (Unit: US$)
1st prize	4,000,000	1/10,000,000	0.4
Numbers before and after the 1st prize	1,000,000	1/5,000,000	0.2
Exact numbers for the 1st prize but the wrong group	1000	1/101,010	0.0099
2nd prize	300,000	1/3,330,000	0.09009
3rd prize	10,000	1/100,000	0.1
4th prize	1000	1/10,000	0.1
5th prize	30	1/100	0.3
6th prize	3	1/10	0.3
Total expected value	–	–	**1.5**

Note Although the actual lottery prizes are in Japanese Yen, the table presents the winning amounts in US dollars for easier understanding. However, due to fluctuations in exchange rates, these values may not accurately reflect the actual prize amounts in Japanese Yen

probability. Based on the table, the expected value of a ticket is approximately $1.5, which means that, on average, for every $3 spent on lottery tickets, players can expect to receive only $1.5 back.

This expected value accounts for the possibility of winning a large prize. However, most participants will win nothing or only a small amount, resulting in an average return significantly lower than the $1.5 expectation. Despite the unfavorable expected value, people continue to buy lottery tickets due to psychological factors such as risk-taking tendencies, the excitement of potential winnings, and the tendency to overestimate small probabilities.

Question: Considering that the expected value of a lottery ticket is lower than its cost, why do people still choose to participate in lotteries?"

St. Petersburg Paradox (Bernoulli's Game): Imagine a game where you repeatedly toss a fair coin until it lands on heads. The payout structure is as follows:

- Heads on the first toss: Win $2.
- Heads on the second toss: Win $4.
- Heads on the third toss: Win $8.
- And so on. If heads appears for the first time on the n-th toss, you win $\$2^n$.

The probability of getting heads for the first time on the n-th toss is $\left(\frac{1}{2}\right)^n$. Therefore, the *expected value* of the game is given by:

$$\text{Expected Value} = \sum_{n=1}^{\infty} \left(\frac{1}{2}\right)^n \times 2^n = \infty$$

Mathematically, this game's expected value is infinite, yet this leads to a paradox in human behavior.

The Paradox: Despite the theoretically infinite expected value, most people would be reluctant to pay a substantial sum to play this game. This reluctance is evident in the hypothetical scenario of a nobleman offering this game to the citizens of St. Petersburg. Few were willing to bet significant amounts despite the potentially infinite winnings, highlighting a discrepancy between mathematical expectation and actual human behavior. This observation led to the term "St. Petersburg Paradox."

The "St. Petersburg Paradox" is named after the city of St. Petersburg, Russia, where eighteenth-century mathematician Daniel Bernoulli lived when he proposed this problem as a thought experiment. This paradox arises because humans do not evaluate decisions based purely on expected value. Instead, they consider factors such as diminishing marginal utility[4] and risk aversion, leading to a reluctance to pay a high entry fee for the game.

Question: In the St. Petersburg Paradox (Bernoulli's Game), why are people unwilling to pay a large sum of money despite the game's theoretically infinite expected value?

1.2.1.3 Beyond Expected Value

The following theories offer deeper insights into the complexities of human decision-making:

A. *Expected Utility Theory*: This theory accounts for individual preferences and risk attitudes, recognizing that the subjective value (utility) of an outcome does not always correspond directly to its monetary value.
B. *Prospect Theory*: Developed by Daniel Kahneman and Amos Tversky, this influential theory enhances our understanding of decision-making under uncertainty by incorporating psychological factors such as loss aversion and framing effects.[5]

These theories provide a more realistic framework for analyzing decision-making in real-world contexts, including financial investments, insurance choices, and gambling behavior. Below, we will explore these theories in more detail.

Expected Utility Theory: Daniel Bernoulli observed that individuals often make decisions not to maximize expected monetary value but rather to maximize the *expected utility*—the subjective value or satisfaction derived from a particular outcome. This concept, which suggests that individuals choose options that maximize their expected utility, was further developed by von Neumann and Morgenstern (1944) in their seminal work on expected utility theory. In this context, *utility* represents a subjective measure of satisfaction or preference.

---

[4] Diminishing marginal utility is when the additional satisfaction or utility gained from consuming an extra unit of a good or service decreases as the quantity consumed increases.

[5] Framing effects occur when people make different decisions based on how the same information is presented, even though the underlying facts remain the same.

Example: Consider a simple coin toss game. If the coin lands on heads, you win \$600; if it lands on tails, you win nothing. The participation fee for this game is \$400.

The expected value is given by:

$$\text{Expected value} = \left(\frac{1}{2} \times 600\right) + \left(\frac{1}{2} \times 0\right) = 300$$

Since the expected value (\$300) is less than the participation fee (\$400), individuals solely focused on maximizing the expected monetary value would not participate.

Now, let us assume an individual's utility for money is proportional to the square of the monetary amount. The expected utility is calculated as follows:

$$\text{Expected utility} = \left(\frac{1}{2} \times 600^2\right) + \left(\frac{1}{2} \times 0^2\right) = 180{,}000$$

Suppose the utility of the participation fee is:

$$400^2 = 160{,}000$$

Then, this individual will participate in the game because the expected utility of the game (180,000) exceeds the utility of the participation fee (160,000).

This example illustrates how individuals may prioritize utility maximization over simple monetary gains, particularly when their utility function is nonlinear (e.g., exhibiting diminishing marginal utility). For instance, expected utility theory is widely applied in finance, insurance, and behavioral economics to model how individuals make investment decisions and assess risks.

While expected utility theory provides a valuable framework for understanding decision-making under uncertainty, it has several limitations:

1. *Deviations from Idealized Utility Maximization*: Real-world decision-making often diverges from the idealized principles of utility maximization. Psychological factors—such as cognitive biases, emotional influences, and social pressures—can significantly impact choices. Moreover, an individual's utility function is not static but rather dynamic, evolving over time due to learning, experiences, and changing circumstances. Additionally, the theory does not fully account for ambiguity aversion, where individuals may prefer known risks over unknown ones.
2. *Individual Differences in Decision-Making*: Individual characteristics play a crucial role in shaping decision-making, which the theory does not fully account for. Emotions such as joy, sadness, fear, and regret can profoundly impact perceived utility. People also exhibit varying risk tolerance levels; some seek out risk, while others are risk-averse. Furthermore, time preference, or the inclination to favor immediate rewards over delayed gratification, significantly influences decision-making.

These limitations underscore the need for more nuanced and comprehensive models that better reflect the complexities of human decision-making under uncertainty.

Prospect Theory: Kahneman and Tversky (1979) developed prospect theory (PT), a foundational concept in behavioral economics that challenges traditional expected utility theory (EUT). While EUT assumes that individuals make decisions to maximize their expected utility, PT demonstrates that human decision-making systematically deviates from this rational model.

According to EUT, individuals:

- Evaluate utility based on final wealth (absolute outcomes).
- Treat gains and losses symmetrically, depending only on utility.
- Assess probabilities objectively.

In contrast, prospect theory assumes that people:

- Evaluate outcomes relative to a reference point (e.g., their current wealth) rather than their final wealth.
- Exhibit loss aversion—losses loom larger than equivalent gains.
- Overweight low probabilities and underweight high probabilities.

Examples: The following examples suggest differences between EUT and PT:

1. *Reference Dependence*: Imagine two individuals, both starting with $1000:
   - Person A receives a gain of $500.
   - Person B also receives a gain of $500, but their initial reference point was $2000 (they had previously lost $1000).

   Under EUT, both individuals end up with $1500, so their utility should be the same. Under PT, however:

   - Person A perceives the $500 as a pure gain, which feels positive.
   - Person B perceives the $500 as a partial recovery from a loss ($2000 → $1500), which feels less favorable.

   Thus, the reference point affects how each person evaluates the outcome, even though their final wealth is identical.
2. *Loss Aversion*: Suppose a person is offered the following choices:
   - Option X: A 50% chance to gain $50 and a 50% chance to lose $50.
   - Option Y: No gain or loss (status quo).

   Under EUT, the gamble has an expected value of $0, so a risk-neutral person would be indifferent between X and Y. Under PT, however, the loss of $50 is psychologically more painful than the equivalent gain of $50 is pleasurable. As a result, many people reject the gamble and prefer Option Y.
3. *Probability Weighting*: Consider the following case: You have a 1% chance of winning $100,000.

- EUT suggests that the decision should be based purely on the expected value of the gamble.
- PT, however, shows that people overweight small probabilities, making this gamble feel more attractive than its actual expected utility.

Here are some real-world implications:

- EUT explains rational behavior but often fails to predict real-world decision-making.
- PT incorporates psychological factors and better explains behaviors such as:
  - Over-insuring against small risks (due to overweighting low probabilities).
  - Rejecting fair bets (due to loss aversion).
  - Purchasing lottery tickets (due to overweighting small probabilities of significant gains).

These differences highlight why prospect theory is widely used in behavioral economics, while expected utility theory remains a cornerstone of classical economics.

1.2.1.4 Alternative Decision-Making Criteria Under Uncertainty

When faced with uncertainty, decision-makers often rely on probabilities to guide their choices. However, when probabilities are uncertain or subjective, alternative criteria can provide valuable insights:

A. ***Laplace Criterion***: Assuming all possible states of the world are equally likely, select the option with the highest average payoff across all states.
B. ***Maximum Likelihood Criterion***: Assuming the most likely state will occur, select the option that yields the highest payoff in that state.
C. ***Aspiration Level Criterion***: Assuming a minimum acceptable level of payoff, select the option that is most likely to meet or exceed this threshold.

These criteria offer different perspectives on decision-making under uncertainty. The choice of criterion depends on factors such as the decision-maker's risk tolerance, available information, and the specific decision context. Let us examine these criteria more closely.

Laplace Criterion: The Laplace criterion, also known as the principle of insufficient reason, is a decision-making rule that assumes equal probability for all possible states of nature when objective probability estimates are unavailable. This principle, attributed to eighteenth-century mathematician Pierre-Simon Laplace, suggests that if there is no reason to believe one outcome is more likely than another, we should assign equal probabilities to all possible outcomes.

Formally, the Laplace criterion selects the option a^* that maximizes the average payoff across all possible states of nature, assuming each state has an equal probability of occurrence:

$$a^* \in \underset{m}{\operatorname{argmax}} \frac{1}{N} \sum_{n=1}^{N} g(a_m, s_n), \quad m \in \{1, 2, \ldots, M\}$$

In Table 1.3, where objective weather probabilities are not available, we can apply the Laplace criterion by assuming equal probabilities for "Sunny," "Cloudy," "Rainy," and "Snowy" (i.e., 0.25 for each).

Since the row-wise sum of payoffs in Table 1.3 is 1000 for all food options, the average payoff (and thus the Laplace criterion value) for each option is:

$$\frac{1000}{4} = 250$$

In this case, the Laplace criterion does not indicate a preference for any particular food option, as all options have the same average payoff.

A practical example of the Laplace criterion can be seen in business decision-making. Suppose a company is choosing a new market to enter but lacks sufficient data on each market's success probability. Using the Laplace criterion, the company assumes that all markets have equal chances of success and selects the one with the highest average expected profit across different scenarios.

Maximum Likelihood Criterion: The maximum likelihood criterion focuses solely on the state of nature with the highest probability of occurrence. It selects the option that yields the highest profit, assuming that the most probable state will occur with certainty.

Let s^* be the state with the highest probability:

$$s^* \in \{s_n | p(s_n) \geq p(s_k), \text{ for all } k \in \{1, 2, \ldots, N\}\}$$

Then, the optimal solution a^* is given by

$$a^* \in \underset{m}{\operatorname{argmax}}\, g\big(a_m, s^*\big), \quad m \in \{1, 2, \ldots, M\}$$

This criterion has a significant limitation: If multiple options yield the same maximum payoff under the most probable state, the criterion fails to provide a unique decision.

Example: Referring back to Table 1.3, given the probabilities:

- Sunny: $p_1 = 0.2$
- Cloudy: $p_2 = 0.5$
- Rainy: $p_3 = 0.3$
- Snowy: $p_4 = 0.0$

Since "Cloudy" has the highest probability ($p_2 = 0.5$), we evaluate the payoffs under cloudy conditions. Both "Fried Noodles" and "French Fries" yield the same profit of \$300. Thus, the maximum likelihood criterion fails to select a unique optimal option between them.

Investors may use the maximum likelihood criterion in financial markets to predict future stock prices. If analysts assume a specific market condition to be the most probable, they may base investment decisions on the expected return of that condition. However, if two or more assets provide the same expected return under the most likely market state, the criterion does not provide a decisive recommendation.

Aspiration Level Criterion: The aspiration level criterion focuses on achieving a specific, predetermined level of performance or outcome rather than maximizing absolute gain. This concept aligns with the work of Herbert Simon (1955), who argued that due to *bounded rationality*, decision-makers often aim to achieve "satisficing" solutions—options that meet a minimum acceptable level of performance—rather than striving for the absolute optimal solution.

Let r represent the aspiration level of the gain. Define the variable b_{mn} as follows:

$$b_{mn} = \begin{cases} 1, & \text{if } g(a_m, s_n) \geq r \\ 0, & \text{if } g(a_m, s_n) < r \end{cases}$$

The options chosen according to the aspiration level criterion are determined by:

$$a^* \in \underset{m}{\operatorname{argmax}} \sum_{n=1}^{N} p(s_n) b_{mn}, \quad m \in \{1, 2, \ldots, M\}$$

Example: Using Table 1.3 and the given probabilities (Sunny: 0.2, Cloudy: 0.5, Rainy: 0.3, Snowy: 0.0):

1. Aspiration Level = \$300:

   - "Fried Noodles" and "French Fries" meet or exceed this aspiration level under "Sunny" and "Cloudy" conditions.
   - Probability of meeting the aspiration level:

$$0.2 \text{ (Sunny)} + 0.5 \text{ (Cloudy)} = 0.7$$

2. Aspiration Level = \$700:

   - "Soft Serve" and "Crepes" meet this higher aspiration level under "Sunny" conditions.
   - The probability of meeting the aspiration level:

$$0.2 \text{ (Sunny)}$$

In the example in Table 1.3, this criterion does not allow for a final decision, as multiple options meet the aspiration level, but no single option dominates.

The aspiration level criterion is widely used in personal finance. For instance, when choosing between job offers, an individual may set a minimum acceptable salary (e.g., $60,000 per year) rather than simply selecting the highest-paying option. Similarly, in business decision-making, a company might choose an investment that guarantees a minimum return of 5% rather than aiming for the highest possible but uncertain return.

1.2.2 Decision-Making Under Strict Uncertainty

This subsection explores decision-making strategies used in situations of strict uncertainty, where the probabilities of potential outcomes are unknown or unknowable. We examine three prominent approaches:

A. *Maximin*: Selects the option that maximizes the minimum possible payoff, favoring a risk-averse approach.
B. *Maximax*: Selects the option that maximizes the maximum possible payoff, favoring a risk-seeking approach.
C. *Minimax Regret*: Selects the option that minimizes the worst-case regret, balancing risk and reward.

These strategies provide distinct perspectives and can lead to different choices depending on the decision-maker's risk tolerance and priorities.

1.2.2.1 Maximin

The Maximin is a conservative decision-making approach that minimizes potential losses. It identifies the state with the lowest payoff for each option and selects the option with the highest payoff among these lowest-payoff states.

Mathematically, the optimal option a^* is given by

$$a^* \in \underset{m}{\operatorname{argmax}} \min_{n} g(a_m, s_n), \quad n \in \{1, 2, \ldots, N\}, m \in \{1, 2, \ldots, M\}$$

Question: To be sure, explain the formula above in words.

Answer: We are searching for the value of a_m (denoted as a^*) that achieves the best worst-case performance of the function $g(a_m, s_n)$. Specifically:

1. For each candidate a_m, evaluate how it performs with respect to all possible s_n.
2. Focus on the worst-case scenario for each a_m by taking the minimum value of $g(a_m, s_n)$ across all $n \in \{1, 2, \ldots, N\}$.
3. Finally, select the a^* that maximizes this worst-case value.

Example: For the cases in Table 1.3, the worst-case state (lowest payoff) for each option is as follows:

- The worst state of "Fried Noodles" is "Snowy", with a payoff of \$0.
- The worst state of "French Fries" occurs in both "Rainy" and "Snowy," resulting in a payoff of \$150.
- The worst state of "Soft Serve" is shared by "Rainy" and "Snowy," with a payoff of \$0.
- The worst state of "Crepes" is "Snowy", with payoff of \$0.

Based on the Maximin criterion, "French Fries" is the optimal choice because it has the highest worst-case payoff (\$150). However, if it were certain that "Snowy" would not occur, the preferred choice would be "Fried Noodles" instead.

The Maximin criterion has several limitations:

- Overly Cautious Approach: It prioritizes minimizing losses, which may lead to missed opportunities for significant gains.
- Ignoring Probabilities: If the probabilities of different states are known, using the Maximin criterion without considering them can result in suboptimal decisions.

Extreme Example: Consider the following two lotteries:

- Lottery A: 99 out of 100 will win \$1000, and 1 out of 100 will win \$100.
- Lottery B: 99 out of 100 will win \$500, and 1 out of 100 will win \$300.

According to the Maximin criterion, we would select Lottery B since its minimum possible gain is \$300, compared to \$100 for Lottery A. However, when calculating the expected value, we see:

$$\text{Expected Value of Lottery}\, A = (0.99 \times 1000) + (0.01 \times 100) = 991$$

$$\text{Expected Value of Lottery}\, B = (0.99 \times 500) + (0.01 \times 300) = 498$$

Since Lottery A has a significantly higher expected value (\$991 vs. \$498), the principle of maximizing expected value would suggest choosing Lottery A. This example illustrates how prioritizing worst-case scenarios may lead to suboptimal decisions when probabilities are considered.

The Maximin criterion is often used in situations where probabilities are uncertain or unknown, such as:

- Game theory, particularly in zero-sum games where an opponent's strategy is unpredictable.
- Competitive bidding, where businesses must prepare for the worst possible market conditions.

1.2.2.2 Maximax

The Maximax is an optimistic decision-making approach that maximizes potential gain. It chooses the option that yields the highest possible payoff without considering the likelihood of that outcome occurring.

Mathematically, the optimal option a^* is given by:

$$a^* \in \underset{m}{\text{argmax}} \max_{n} g(a_m, s_n), \quad n \in \{1, 2, \ldots, N\},\ m \in \{1, 2, \ldots, M\}$$

Example: For the cases in Table 1.3, the best-case state (highest payoff) for each option is as follows:

- The best state of "Fried Noodles" is "Sunny", with a payoff of $500.
- The best state of "French Fries" is "Sunny", with a payoff of $400.
- The best state of "Soft Serve" is "Sunny", with a payoff of $800.
- The best state of "Crepes" is "Sunny", with a payoff of $700.

Since "Soft Serve" has the highest possible payoff ($800), it is the optimal choice based on the Maximax criterion.

This approach prioritizes the potential for the highest possible gain, even at the risk of significantly low or no payoff:

- "Soft Serve" is the best choice if it is sunny but the worst in other weather conditions.
- Selecting "Soft Serve" reflects a high tolerance for risk.

The Maximax criterion can be overly optimistic, focusing solely on the best-case scenario and ignoring potential losses. This criterion is applicable when a high-risk, high-reward approach is desired. However, before making a final decision, it is essential to consider its limitations and evaluate other factors, such as potential losses and risk tolerance.

Maximax vs. Maximin—A Diversified Investment Example: To illustrate the differences between Maximax and Maximin, consider the following example. Imagine two friends have asked you to provide financial support for the school festival:

- Mr. A proposes a concert at the playground.
- Mr. B suggests hosting a movie party in the gymnasium.

Suppose you are thinking about investing $1000. The investment options are as follows:

- Mr. A (Concert)
  - Under sunny or cloudy conditions: 100% profit ($2000 return on a $1000 investment)

  - Under rainy or snowy conditions: 80% loss ($200 return on a $1000 investment)
- Mr. B (Movie Party)
  - Under sunny or cloudy conditions: 50% loss ($500 return on a $1000 investment)
  - Under rainy or snowy conditions: 70% profit ($1700 return on a $1000 investment)

Now, let us compare how Maximax and Maximin approaches influence investment decisions.

- Maximax Approach
  - Mr. A: Maximum gain is 100% profit (+$1000)
  - Mr. B: Maximum gain is 70% profit (+$700)
  - Decision: Allocate all $1000 to Mr. A's concert, as it offers the highest potential profit.
- Maximin Approach
  - Mr. A: Minimum gain is 80% loss (−$800)
  - Mr. B: Minimum gain is 50% loss (−$500)
  - Decision: Select to invest all $1000 in Mr. B's movie party to mitigate potential losses.

The choice depends heavily on the investor's risk tolerance. Financial goals, investment time horizons, and personal circumstances significantly influence risk preference.

- Without probability information, the Maximin approach (investing in Mr. B) offers a more conservative strategy to minimize potential losses.
- If the investor is willing to take on more risk, the Maximax approach (investing in Mr. A) could yield higher returns.

Diversification Strategy: Alternatively, consider diversifying your investment in $100 increments. Table 1.6 illustrates the profit/loss for various combination investments in the concert and movie party under "sunny or cloudy" and "rainy or snowy" conditions. The rightmost columns show the maximum and minimum profits across these two weather conditions.

Below are explanations of the columns in Table 1.6:

- Investment: Amount invested in the concert (first value) and movie party (second value), in $10 units (e.g., "40 | 60" means $400 in the concert and $600 in the movie party).
- Profit (sunny/cloudy) (Concert): Individual profit from the concert investment in each scenario when the weather is sunny or cloudy.
- Profit (sunny/cloudy) (Movie): Individual profit from the movie party investment in each scenario when the weather is sunny or cloudy.

Table 1.6 Diversified investment and profit (Unit: $10)

Investment		Profit (sunny/cloudy)			Profit (rainy/snowy)			Profit	
Concert	Movie	Concert +100%	Movie −50%	Total	Concert −80%	Movie +70%	Total	Max	Mini
0	100	0	−50	**−50**	0	70	**70**	70	−50
10	90	10	−45	**−35**	−8	63	**55**	55	−35
20	80	20	−40	**−20**	−16	56	**40**	40	−20
30	70	30	−35	**−5**	−24	49	**25**	25	−5
40	60	40	−30	**10**	−32	42	**10**	10	10
50	50	50	−25	**25**	−40	35	**−5**	25	−5
60	40	60	−20	**40**	−48	28	**−20**	40	−20
70	30	70	−15	**55**	−56	21	**−35**	55	−35
80	20	80	−10	**70**	−64	14	**−50**	70	−50
90	10	90	−5	**85**	−72	7	**−65**	85	−65
100	0	100	0	**100**	−80	0	**−80**	100	−80

- Profit (sunny/cloudy) (Total): Total profit for each investment allocation when the weather is sunny or cloudy.
- Profit (rainy/snowy) (Concert): Individual profit from the concert investment in each scenario when the weather is rainy or snowy.
- Profit (rainy/snowy) (Movie): Individual profit from the movie party investment in each scenario when the weather is rainy or snowy
- Profit (rainy/snowy) (Total): Total profit for each investment allocation when the weather is rainy or snowy.
- Profit (Max): Maximum profit across both weather conditions for each investment allocation.
- Profit (Mini): Minimum profit across both weather conditions for each investment allocation.

Using the Maximax and Maximin strategies, you will get the following results respectively:

- Maximax (Look at the "Max" Column)
  - The row where Investment = 100 | 0 (investing all $1000 in the concert) shows the Maximax profit of $1000.
  - However, if the weather is rainy or snowy, the loss is significant (-$800).
- Maximin (Look at the "Min" Column)
  - The row where Investment = 40 | 60 (investing $400 in the concert and $600 in the movie party) shows the Maximin profit of $100.
  - This investment will give you a profit of $100 in any weather.

Next, we will explore an approach that seeks to balance the strategies of Maximax and Maximin.

Hurwicz Criterion—Balancing Maximax and Maximin: Hurwicz (1951) introduced a decision criterion that combines the optimistic Maximax and the pessimistic Maximin approaches, striking a balance between the two. This balance is determined by the *coefficient of optimism*, α, a numerical value between 0 (pure pessimism) and 1 (pure optimism).

- Optimistic decision-makers (higher α) prioritize the potential for the best possible outcomes (Maximax criterion).
- Pessimistic decision-makers (lower α) prioritize avoiding the worst outcomes (Maximin criterion).

The Hurwicz criterion reduces to the Maximax criterion when $\alpha = 1$ and the Maximin criterion when $\alpha = 0$.

For each option a_m, the criterion calculates a weighted average of the maximum and minimum payoff, using α and $1 - \alpha$ as weights, respectively. The option with the highest weighted value, a^*, is chosen:

$$a^* \in \underset{m}{\operatorname{argmax}} \left\{ \alpha \times \max_n g(a_m, s_n) + (1 - \alpha) \times \min_n g(a_m, s_n) \right\}$$

where $n \in \{1, 2, \ldots, N\}$ and $m \in \{1, 2, \ldots, M\}$.

Example: To illustrate the Hurwicz criterion, consider Table 1.3. Since the "Snowy" case offers a zero payoff for three options, it is excluded from the calculations.

For each option, the weighted maximum and minimum payoffs (based on α and $1 - \alpha$) are as follows:

- Fried Noodles: $\alpha \times 500 + (1 - \alpha) \times 200 = 300\alpha + 200$
- French Fries: $\alpha \times 400 + (1 - \alpha) \times 150 = 250\alpha + 150$
- Soft Serve: $\alpha \times 800 + (1 - \alpha) \times 0 = 800\alpha$
- Crepes: $\alpha \times 700 + (1 - \alpha) \times 100 = 600\alpha + 100$

For any value of $0 \leq \alpha \leq 1$, the weighted value of "Fried Noodles" exceeds that of "French Fries." Thus, "French Fries" is never selected under the Hurwicz criterion.

Recall the following:

- "French Fries" was selected based on the Maximin criterion, but when "Snowy" was excluded, "Fried Noodles" became the preferred option.
- "Soft Serve" was chosen under the Maximax criterion.

From the calculations:

- "Fried Noodles" is selected when $0 \leq \alpha \leq \frac{1}{3}$, reflecting more pessimistic decision-making.
- "Crepes" is selected when $\frac{1}{3} \leq \alpha \leq \frac{1}{2}$, indicating moderate optimism.

- "Soft Serve" is selected when $\frac{1}{2} \leq \alpha \leq 1$, appealing to optimistic decision-makers.

The Hurwicz criterion forms a crucial axiomatic foundation for decision-making under uncertainty. Foundational works by Arrow (1951), Savage (1954), and Luce and Raiffa (1957) provide additional insights into its theoretical and philosophical basis. These works explore the mathematical and philosophical principles underlying decision-making under uncertainty.

1.2.2.3 Minimax Regret

The Minimax Regret is a decision-making approach aimed at minimizing the feeling of loss or missed opportunity, defined as the difference between:

- The payoff of the best possible decision and
- The payoff of the chosen decision, given a specific state of nature.

This criterion, also known as the Savage criterion, helps choose the option that minimizes the maximum regret across all possible states. Opportunity loss refers to the benefit forgone by not making the optimal decision. It is a central concept in decision theory with broad applications, as it quantifies the cost of a suboptimal choice.

The regret associated with choosing a_m in a realized state s_n is mathematically defined as:

$$r(a_m, s_n) = \max_k g(a_k, s_n) - g(a_m, s_n), \quad k \in \{1, 2, \ldots, M\}$$

where

- $g(a_k, s_n)$ is the payoff when option a_k is chosen and state s_n occurs.
- $r(a_m, s_n)$ is the regret of choosing a_m when s_n is the realized state, calculated as the difference between the best possible pay off (across all k) and the payoff for a_m.

This formula quantifies the regret for each decision and serves as the basis for identifying the decision with the smallest possible regret in the worst-case scenario.

The Minimax Regret criterion selects the option a^* that minimizes the maximum regret across all possible states of nature s_n:

$$a^* \in \underset{m}{\operatorname{argmin}} \max_n r(a_m, s_n), \quad n \in \{1, 2, \ldots, N\}, \; m \in \{1, 2, \ldots, M\}$$

Example: Consider again Table 1.3. The maximum regret for each option is calculated as follows:

- Fried Noodles: \$300
- French Fries: \$400

- Soft Serve: $200
- Crepes: $150

Under the Minimax Regret criterion, "Crepes" is selected as the optimal choice due to its lowest maximum regret of $150.

The calculation of regret values can follow the systematic approach described in Table 1.7. The regret matrix summarizes opportunity losses for each decision option across all states of nature.

Steps to create a regret matrix are as follows:

1. Identify the Maximum Payoff for Each State: Determine the best possible payoff for each state of nature in the Payoff Matrix.
   - Sunny: (800) (Soft Serve)
   - Cloudy: (300) (Fried Noodles, French Fries)
   - Rainy: (200) (Fried Noodles)
   - Snowy: (150) (French Fries)
2. Calculate Regret for Each Cell: Subtract each cell's payoff from the maximum payoff in the same column.
   - Example: French Fries (Rainy) = (200 − 150 = 50)
3. Determine the Maximum Regret for Each Option: For each decision option, find the largest regret across all states.
   - Example: Soft Serve: Max regret (200)
4. Select the Option with the minimum Maximum Regret: Select the option with the smallest value in the last column.
   - In this case, "Crepes" has the smallest max regret (150), making it the best choice.

Table 1.7 A regret (opportunity loss) matrix for finding the minimum regret option

Option	Payoff matrix (Table 1.3)				Regret matrix (Dollars)				Maximum regret (Dollars)
	Sunny	Cloudy	Rainy	Snowy	Sunny	Cloudy	Rainy	Snowy	
Fried Noodles	500	300	200	0	300	0	0	150	300
French Fries	400	300	150	150	400	0	50	0	400
Soft Serve	800	200	0	0	0	100	200	150	200
Crepes	700	200	100	0	100	100	100	150	150

The Minimax Regret criterion has some limitations:

- Overly Conservative: This approach avoids the worst regret, but it may overlook opportunities for greater rewards in more favorable scenarios. For example, in high-stakes investment scenarios, this approach might overlook opportunities for significant gains in favor of minimizing losses.
- Neglects Probabilities: It does not account for probability information, which may lead to suboptimal decisions when probabilities are available.

Savage (1954) argued that the Minimax Regret criterion is a rational decision-making framework because:

1. Robustness to Uncertainty: It minimizes the worst possible regret, making it effective in uncertain situations.
2. Ease of Use: The regret matrix provides a structured and systematic approach, enabling straightforward calculations for even complex decision-making problems.

Since Savage's work, researchers have extensively studied and applied the Minimax Regret criterion across various fields. This criterion's ability to minimize worst-case regret has proven valuable in numerous domains, such as:

- Economics: The Minimax Regret criterion helps policymakers design robust policies that account for uncertainties and minimize the potential for worst-case outcomes. For instance, it is applied in trade agreements or taxation strategies to prevent significant societal losses under unexpected scenarios.
- Finance: Financial analysts use the criterion in investment strategies to balance risk and return. By focusing on avoiding extreme regret, investors can make decisions that minimize potential losses during market downturns while still pursuing reasonable gains.
- Information Theory: The Minimax Regret criterion provides a structured approach to optimizing information efficiency in data compression and decision-making under uncertainty. It helps quantify and minimize the loss incurred by suboptimal choices in encoding or transmitting data.
- Artificial Intelligence: AI researchers leverage the criterion to minimize worst-case losses in uncertain or dynamic environments. For example, the Minimax Regret principle is applied in training autonomous vehicles to prioritize safety in unpredictable traffic situations.
- Game Theory: In competitive scenarios, the Minimax Regret criterion is used to develop strategies that minimize potential losses against unknown or adversarial opponents. It provides a systematic way to prepare for the worst-case responses from an adversary while optimizing overall performance.

1.3 Game Theory for Strategic Decision-Making

Strategic decision-making involves scenarios where outcomes are influenced not only by one's own choices but also by the actions and decisions of competitors. Such decisions fall within the broader category of decision-making under uncertainty. Game theory, established by von Neumann and Morgenstern (1944), provides a mathematical framework for analyzing such situations.

Initially, game theory focused on two-person zero-sum games, characterized by the following features:

- Two-person games: Scenarios involving only two players.
- Zero-sum games: Situations where one player's gain equals the other player's loss.

Such games highlight the competitive nature of decision-making, where one player's success directly implies the other's failure. A sports competition between two teams, where one team's victory results in the other's defeat, is a typical example of a two-person zero-sum game.

While zero-sum games focus on pure competition, later research expanded to non-zero-sum games, which account for scenarios where participants' total gains and losses do not necessarily balance out. Notable contributions, such as Nash's equilibrium theory (Nash, 1951), advanced the study of these more complex and realistic interactions. For example, while trade negotiations theoretically represent a non-zero-sum game where mutual benefits are possible, real-world disputes often arise due to uneven distribution of gains, conflicting domestic priorities, or regulatory barriers.

1.3.1 Strategic-Form Games

In game theory, the participants—known as players—engage in situations referred to as games, where the focus of study lies in their decision-making processes. Game theory can be broadly divided into two types:

- Cooperative games: Players work together, considering each other's decisions to achieve mutually beneficial outcomes.
- Non-cooperative games: Players act independently, often pursuing strategies that maximize their own payoffs without collaborating with others.

This discussion focuses on non-cooperative games, which are further classified into two forms:

- Strategic-form games (or normal-form games): Players choose their strategies simultaneously, each considering the potential choices of others.

- Extensive-form games: Players select their strategies sequentially, taking into account the timing and order of moves. An example of an extensive-form game is chess, where players take turns making moves in a well-defined sequence.

This discussion focuses on strategic-form games characterized by simultaneous decision-making, where players choose their strategies without knowledge of others' choices. The combination of strategies all players select determines the game's outcome and the corresponding payoffs for each player.

Key Assumption: In this context, it is generally assumed that players act rationally, aiming to maximize their expected utility or payoff. This assumption underpins the analysis of strategic-form games, as it ensures that players consider all available information and potential outcomes to optimize their decisions.

Example—Payoff Matrix in a Strategic-Form Game: In strategic-form games, a payoff matrix is a useful tool to illustrate the interactions between two players with multiple strategies. It is a table that describes the payoffs for each player based on their chosen strategies. Table 1.8 provides an example of such a payoff matrix.

The payoff matrix shows that both players have three strategies available. Each entry contains two numbers: the first represents Player 1's payoff, and the second represents Player 2's payoff. The order of the numbers is consistent throughout the matrix.

For instance, if Player 1 chooses Strategy C and Player 2 selects Strategy D, the resulting payoffs are +3 for Player 1 and −3 for Player 2. This outcome highlights the zero-sum nature of the game, where one player's gain exactly offsets the other's loss.

Using the payoff matrix, players can engage in strategic reasoning:

1. Player 1 might initially consider Strategy A, aiming for the maximum profit of 4 points.
2. Player 2, anticipating this, might choose Strategy F to gain 2 points instead of losing 4 points with Strategy D.
3. Player 1, in response, might switch to Strategy B, seeking a 3-point gain.
4. Player 2, expecting this, might counter with Strategy D.

This cycle of anticipation and counter-anticipation showcases the complexity of decision-making in such games, often preventing players from reaching a stable outcome.

Table 1.8 Example of a payoff matrix (Kojima, 2013)

		Player 2		
		Strategy D	Strategy E	Strategy F
Player 1	Strategy A	+4, −4	−1, +1	−2, +2
	Strategy B	−3, +3	±0, ±0	+3, −3
	Strategy C	+3, −3	+1, −1	+2, −2

Table 1.9 Minimum gains for each strategy (Player 1's payoff)

		Player 2			Minimum gain of Player 1
		Strategy D	Strategy E	Strategy F	
Player 1	Strategy A	4	−1	−2	−2
	Strategy B	−3	0	3	−3
	Strategy C	3	1	2	1
Minimum gain of Player 2		4	1	3	

To break this cycle, players can evaluate each strategy's minimum guaranteed payoff—the least amount they can ensure, regardless of the opponent's choice. This idea leads to the *Maximin strategy*, where players aim to maximize their minimum possible payoff.

Table 1.9 adds one row and column to Table 1.8, specifying the minimum gain for each player's strategies. Since this is a zero-sum game, Player 2's payoffs can be determined by inverting the signs of Player 1's payoffs.

Table 1.9 suggests that:

- Player 1 might choose Strategy C, which guarantees a minimum gain of 1 point. Choosing Strategy B would risk a loss of 3 points.
- Player 2, seeking to minimize their maximum loss, might choose Strategy E, which limits their loss to 1 point.

This strategy combination—Player 1 choosing Strategy C and Player 2 choosing Strategy E—represents a Nash equilibrium. In this state, neither player has an incentive to change their strategy unilaterally. Notably, the Nash equilibrium favors Player 1, as they ensure a minimum gain of 1 point.

We will explore Nash equilibrium in greater detail later.

1.3.2 Prisoners' Dilemma and Nash Equilibrium

The Prisoner's Dilemma is a classic example in game theory, showcasing the tension between cooperation and betrayal, as well as the conflict between individual and collective interests.

Scenario: Two prisoners, A and B, are suspected of jointly committing a crime. However, due to insufficient evidence, the police present them with the following conditions while keeping them in separate rooms:

- If both remain silent (cooperate), each serves 1 year in prison.
- If only one confesses (betrays), the confessor is released, while the silent prisoner serves 10 years.
- If both confess, each serves 5 years in prison.

The Prisoner's Dilemma highlights several essential features:

- *Non-Zero-Sum Game*: The total payoff for both players does not always equal zero, allowing for outcomes where both players can gain, both can lose, or one can gain at the other's expense.
- *Cooperative and Non-Cooperative Elements*: Cooperation (staying silent) results in a minimal sentence for both. However, the temptation to betray (for a chance at freedom) creates a strong incentive to defect.
- *Imperfect Information*: Each prisoner must decide without knowing the other's choice, introducing uncertainty into the decision-making process.

Table 1.10 shows the payoff matrix for the Prisoner's Dilemma, where the first number in each cell represents Prisoner A's sentence, while the second represents Prisoner B's sentence. In this matrix, lower numbers indicate a shorter prison sentence, making them more favorable outcomes for the prisoners.

Prisoner A's reasoning might proceed as follows:

1. If I remain silent (cooperate):
    - If Prisoner B also remains silent, we both serve 1 year in prison.
    - If Prisoner B confesses, I will serve 10 years, which is my worst possible outcome.
2. If I confess (betray):
    - If Prisoner B remains silent, I am released, while Prisoner B serves 10 years.
    - If Prisoner B also confesses, we both serve 5 years.

Given the risks associated with staying silent, confessing (betraying) emerges as the safer option for Prisoner A. Since the game is symmetrical, Prisoner B is likely to reach the same conclusion.

The Dilemma: The Prisoner's Dilemma illustrates a fundamental social conflict:

- If both players act rationally to minimize their own risk, they both end up in a worse situation (5 years each).
- If both cooperate (remain silent), they will serve only 1 year each, which is better collectively but risky individually.
- The fear of exploitation (being betrayed) leads both prisoners to confess, resulting in a suboptimal outcome for both.

The Prisoner's Dilemma is widely used to study cooperation, competition, and decision-making in various fields. For example:

Table 1.10 Payoff matrix of the prisoner's dilemma

		Prisoner B	
		Silence (Cooperation)	Confession (Betrayal)
Prisoner A	Silence (Cooperation)	1 year, 1 year	10 years, free
	Confession (Betrayal)	Free, 10 years	5 years, 5 years

- Economics: Competing firms may engage in price wars to attract customers, even though maintaining higher prices would result in mutual profit.
- Political Science: Arms races between rival nations demonstrate the temptation to prioritize military buildup over disarmament despite the collective benefits of reduced spending.
- Biology: In the evolution of species, cooperative behavior often arises, but it also presents a similar challenge—individuals prioritize their own survival, which can sometimes contradict what benefits the group as a whole.

This model provides a framework for understanding how individual incentives can conflict with group interests and the challenges of fostering cooperation in competitive environments.

Nash Equilibrium: Introduced by John Nash (1950), Nash equilibrium is a cornerstone of non-cooperative game theory. It describes a situation where players act independently without forming binding agreements.

A Nash equilibrium is a set of strategies where no player can unilaterally deviate from their selected strategy to achieve a higher payoff, given the other players' strategies. In simpler terms, a Nash equilibrium occurs when no player can improve their outcome by unilaterally changing their strategy while the other players' strategies remain unchanged.

Definition: For a given player,

- If their opponent chooses strategy (Y) and they maximize their payoff by choosing strategy (X), then (X) is the optimal response to (Y).
- A set of strategies constitutes a Nash equilibrium if, for each player, their chosen strategy is the optimal response to the strategies of all other players.

Characteristics of Nash equilibrium are as follows:

1. *Not Necessarily Optimal Overall*: A Nash equilibrium does not always produce the highest possible combined payoff for all players. For instance, in the Prisoner's Dilemma, the Nash equilibrium (both players confessing) leads to a worse outcome than the cooperative outcome (both remaining silent).
2. *Stability*: A set of strategies that remains stable is considered a Nash equilibrium. However, not all Nash equilibria are guaranteed to occur in practice.

Some games have multiple Nash equilibria, meaning different stable strategy sets may emerge depending on various factors such as:

- Player Expectations: Players may anticipate specific strategies from others based on their reasoning or shared understanding.
- Past Experiences: Historical interactions between players can guide strategy selection.
- Communication Strategies: Indirect signaling or pre-game communication may influence equilibrium selection, though binding agreements are not allowed in non-cooperative games.

Example—Finding a Nash Equilibrium: Two groups have set up "Fried Noodles" stalls at a school festival. On the first day, both groups achieved the same sales. The next day, each group considers lowering the selling price of their noodles to increase profits. However, neither group knows the other's decision in advance.

To simplify, we assume the following:

- If a group lowers its price, the reduction is the same for both groups.
- Today's profits for both groups are 10 units, where "units" represent monetary profits (e.g., 100 dollars per unit).

The potential outcomes are:

- If both groups maintain their prices, they each keep their current profit of 10 units.
- If both groups lower their prices, their profits decrease to 7 units each.
- If one group lowers its price while the other maintains its price, the group that lowers its price increases its profit to 12 units, while the group that maintains its price sees its profit drop to 4 units.

Table 1.11 shows the payoff matrix for this situation. In this matrix, higher numbers indicate greater profits, making them the more favorable outcomes for each group.

Let us analyze the decision-making process for each group:

- Group A:
  - If Group B maintains its price, Group A can increase its profits by lowering its price (option (c)).
  - If Group B lowers its price, Group A will also be better off lowering its price (option (d)) to avoid a significant profit drop.
- Group B: Because the game is symmetrical—where both groups face identical choices and payoffs—Group B will likely reach the same conclusions as Group A.

The Nash equilibrium occurs at (d), where both groups select to reduce their prices. In this situation, neither group has an incentive to unilaterally change its strategy, given the other group's chosen strategy.

Social Dilemma: This scenario highlights a common social dilemma: individual self-interest, while rational in isolation, can lead to collectively worse outcomes when cooperation is needed for optimal results. While both groups would achieve higher profits by maintaining their prices (option (a)), the fear of losing market share to the

Table 1.11 Payoff matrix of price competition

		Group B	
		Price maintenance	Price reduction
Group A	Price maintenance	(a) 10, 10	(b) 4, 12
	Price reduction	(c) 12, 4	(d) 7, 7

competitor motivates each group to lower its prices, ultimately reducing earnings for both.

This simple example demonstrates how rational decisions made by individuals in isolation can lead to collectively worse outcomes.

1.4 Multi-objective Decision-Making

Many real-world decision-making problems require optimizing multiple objectives simultaneously. For example, in business management, it is necessary to balance employee satisfaction and social contribution while maximizing profits. These objectives may conflict, making it challenging to find an optimal solution. The key issue, then, is how to manage trade-offs when achieving one goal requires sacrificing another.

Multi-objective decision-making problems are characterized by the following:

- *Pareto-Optimal Solutions*: A Pareto-optimal solution is one in which no other solution can improve one objective without worsening at least one other. In other words, any attempt to enhance one objective inevitably leads to the deterioration of another. A multi-objective decision problem typically results in a set of Pareto-optimal solutions. Several methods can generate these solutions, including the weighted sum method, lexicographical ordering, constraint handling, and the Maximin approach.
- *Decision-Maker's Preferences*: The decision-maker's preferences ultimately determine the final choice from the Pareto set. These preferences are often expressed through techniques such as utility functions, which quantify the value or satisfaction derived from different outcomes. Methods for assisting a decision-maker in selecting a solution include goal programming and the analytic hierarchy process (AHP). The decision-maker determines the most suitable solution by weighing the importance of each objective and considering the trade-offs between them.

Italian economist Vilfredo Pareto introduced the concept of Pareto optimality in his economic research during the late nineteenth and early twentieth centuries. Pareto optimality describes a state in which, during resource allocation, one person's utility (satisfaction) cannot be increased without decreasing another person's utility. This concept is also referred to as Pareto efficiency.

1.4.1 Pareto-Optimal Solutions

Suppose there are M options, $a_1, a_2, \ldots, a_M$, and N objectives, $s_1, s, \ldots, s_N$. Let $g(a_m, s_n)$ represent the gain of option a_m for objective s_n.

Note: The symbol for "objective" here is the same as that used for the "state" in decision-making under uncertainty. Please take care not to confuse the two.

The concepts of domination are defined as follows:

- *Domination*: Option a_i is said to dominate a_j if a_i is strictly better than a_j in all objectives:

$$g(a_i, s_n) > g\left(a_j, s_n\right), \quad \text{for all } s_n$$

- *Weak Domination*: Option a_i is said to weakly dominate a_j if a_i is at least as good as a_j in all objectives:

$$g(a_i, s_n) \geq g\left(a_j, s_n\right), \quad \text{for all } s_n$$

  and there exists at least one objective s_k such that:

$$g(a_i, s_k) > g\left(a_j, s_k\right)$$

In multi-objective decision-making, selecting an option that is dominated (or weakly dominated) by another is considered irrational. Consequently, the focus is placed on Pareto-optimal and weak Pareto-optimal options.

The concepts of Pareto optimality are described as follows:

- *Pareto Optimal*: An option a_m is Pareto optimal if no other option weakly dominates a_m.
- *Weakly Pareto Optimal:* An option a_m is weakly Pareto optimal if no other option dominates a_m.

Options excluded from the optimal group are categorized as follows:

- Pareto Optimality: This excludes both dominated and weakly dominated options. A Pareto-optimal solution cannot even be weakly dominated by another option.
- Weak Pareto Optimality: This relaxes the requirement of strict dominance, excluding only strictly dominated options. Weakly dominated options are still considered optimal in this case.

1.4.1.1 Example—Which Car Should You Buy?

Imagine you are buying a car, with price and fuel efficiency as the key factors:

- Car A: Costs $20,000 and gets 30 miles per gallon (mpg).
- Car B: Costs $22,000 and gets 35 mpg.
- Car C: Costs $22,000 and gets 30 mpg.

1. Dominance Relationships

- Car A vs. Car B: Car A is better in price ($20,000 vs. $22,000) but worse in fuel efficiency (30 mpg vs. 35 mpg). Since each car is better in one objective, neither dominates the other.
- Car A vs. Car C: Car A weakly dominates Car C because Car A is cheaper ($20,000 vs. $22,000) while having the same fuel efficiency (30 mpg).
- Car B vs. Car C: Car B weakly dominates Car C because Car B is strictly better in fuel efficiency (35 mpg vs. 30 mpg) while having the same price ($22,000).

2. Pareto Optimality

   - Car A is Pareto optimal because no other car weakly dominates it.
   - Car B is Pareto optimal because no other car weakly dominates it.
   - Car C is not Pareto optimal because it is weakly dominated by both Car A (better price, equal fuel efficiency) and Car B (better fuel efficiency, equal price).

3. Weak Pareto Optimality

   - Car A is weakly Pareto optimal because no other car strictly dominates it.
   - Car B is weakly Pareto optimal because no other car strictly dominates it.
   - Car C is also weakly Pareto optimal because no car strictly dominates it, even though it is weakly dominated by both Car A and Car B.

Which Car Should You Buy?

- If price matters most: Select Car A, as it is the cheapest.
- If fuel efficiency matters most: Select Car B, as it has the highest mpg.
- Avoid Car C: It offers no unique advantage compared to Cars A and B.

Note: Pareto optimality refers to a state in which improving one objective (e.g., price or fuel efficiency) requires sacrificing another. In this case, reducing the price means sacrificing fuel efficiency, and improving fuel efficiency means accepting a higher price.

1.4.1.2 Generating Pareto-Optimal Solutions

There are several methods for generating a set of Pareto-optimal solutions:

A. Weighted Sum Method
B. Lexicographical Ordering Method
C. Constraint Method
D. Maximin Method

Below are brief explanations of each method.

Weighted Sum Method: The Weighted Sum Method assigns weights to each objective to reflect their relative importance to the decision-maker.

Let the weight of the objective s_n be $w_n (n = 1, 2, \ldots, N)$, where

$$0 \le w_n \le 1 \quad \text{for all } n, \quad \sum_{n=1}^{N} w_n = 1$$

The weighted sum of the gains for each option a_m is calculated as:

$$\sum_{n=1}^{N} w_n g(a_m, s_n)$$

The optimal option a^* is selected by maximizing the weighted sum:

$$a^* \in \underset{m}{\operatorname{argmax}} \sum_{n=1}^{N} w_n g(a_m, s_n), \quad m \in \{1, 2, \ldots, M\}$$

This method has the following properties, which are well established in multi-objective optimization theory. While detailed proofs are omitted here for brevity, the following descriptions provide an overview of the method's key characteristics:

- It always selects weakly Pareto-optimal options, regardless of the assigned weights.
- If all weights are strictly positive, this method guarantees selecting Pareto-optimal options.
- However, regardless of the weight configuration, this method cannot generate all Pareto-optimal options, meaning some may remain unselected.

Lexicographical Ordering Method: This method involves prioritizing all objectives to ensure there are no ties and then selecting options using the following steps:

1. *First Priority*: Select the option with the greatest gain for the highest-priority objective. If multiple options remain, proceed to the next step.
2. *Second Priority*: Of the options chosen in Step 1, select the one with the greatest gain for the second highest-priority objective. If multiple options remain, proceed to the next step.
3. *Subsequent Priorities*: Continue this process, evaluating objectives in order of priority, until only one option or a set of equally preferable options remains.

This method always selects a Pareto-optimal option. However, some Pareto-optimal options may never be chosen, regardless of the priority order.

Constraint Method: The Constraint Method involves setting minimum acceptable performance thresholds for all but one objective. The method then selects the option that maximizes the gain for the remaining objective while satisfying the predefined constraints on the others.

Let s_k be the objective to maximize. Introduce a threshold t_n for each objective s_n (where $n \neq k$):

$$g(a_m, s_n) \geq t_n, \quad \text{for all } n \neq k$$

Find the optimal option a^* by solving:

$$a^* \in \underset{m}{\text{argmax}}\, g(a_m, s_k), \quad m \in \{1, 2, \ldots, M\}$$

subject to:

$$g(a_m, s_n) \geq t_n, \quad \text{for all } n \neq k$$

The Constraint Method always selects a weakly Pareto-optimal option. If only one option is selected, it is always a Pareto-optimal option. This method can explore different Pareto-optimal solutions by adjusting the constraints and their corresponding thresholds.

Maximin Method: The Maximin Method focuses on selecting the option that maximizes the worst-case performance across all objectives.

For each option a_m, find the minimum gain across all objectives:

$$\min_{n} g(a_m, s_n), \quad n \in \{1, 2, \ldots, N\}$$

Then, select the option a^* that maximizes these minimum gains:

$$a^* \in \underset{m}{\text{argmax}}\, \min_{n} g(a_m, s_n), \quad m \in \{1, 2, \ldots, M\}$$

The Maximin Method always selects a weakly Pareto-optimal option. If the solution is unique, it is guaranteed to be Pareto optimal.

1.4.2 Analytic Hierarchy Process

The Analytic Hierarchy Process (AHP) is a structured technique for organizing and analyzing complex decision-making problems. Developed by Thomas L. Saaty (1980), the AHP is a multi-criteria decision analysis (MCDA) method that is particularly useful in situations where:

- Objective evaluations are challenging: Unlike methods such as goal programming, which rely on precise target values for each objective, the AHP accommodates situations in which quantifying the benefits of each alternative for each criterion is difficult.
- Prioritizing objectives is crucial: the AHP effectively determines the relative importance of different criteria in decision-making.

Note: In the Analytic Hierarchy Process, "options" are referred to as "alternatives," and "objectives" are referred to as "criteria."

1.4.2.1 Key Steps in the AHP

1. *Problem Definition*: Clearly define the overall decision-making goal and break it down into a hierarchy of sub-goals, criteria, and sub-criteria as needed.
2. *Hierarchy Construction*: Organize the decision elements into a hierarchical structure:
   - The goal is at the top (apex),
   - Criteria and, if necessary, sub-criteria in the middle,
   - Alternatives at the bottom.
3. *Pairwise Comparisons*: Systematically compare elements within each hierarchy level in pairs. Assign a numerical value to each pair to represent their relative importance using a predefined scale (e.g., Saaty's 1–9 scale).
4. *Weight Calculation*: Derive the weights or relative priorities of criteria and alternatives from the pairwise comparisons using mathematical techniques such as the Eigenvector Method.
5. *Consistency Checks*: Conduct a consistency check on the pairwise comparisons using the Consistency Ratio (CR). If the CR exceeds an acceptable threshold (typically 0.1), revise the comparisons to improve logical coherence.
6. *Synthesis*: Integrate the weights of criteria and alternatives to compute an overall ranking or preference for the available options. This step combines the relative importance of criteria and alternatives to support decision-making.

1.4.2.2 Understanding the AHP with an Example: Selecting a Food Stall

To illustrate the AHP, consider the decision-making process of selecting a food stall for a school festival. While numerous criteria could be considered—such as profitability, popularity, ease of preparation, ingredient availability, safety, and originality—we will focus on three key factors for simplicity: profitability, popularity, and ease of preparation.

By applying the AHP methodology, decision-makers can systematically evaluate each food stall option based on these criteria, determine the relative importance of each criterion, and ultimately make an informed and well-justified selection.

Constructing the Hierarchy: The first step in the AHP is to construct a hierarchical representation of the decision problem. This hierarchy typically consists of three levels:

1. *Goal*: This represents the overall objective of the decision-making process. In our example, the goal is "Selecting the best food stall for the school festival."

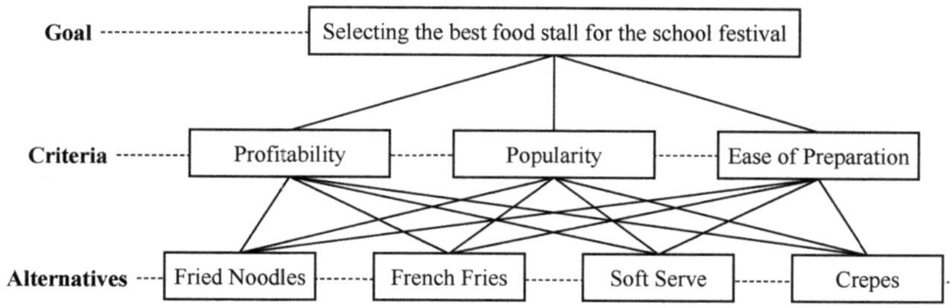

Fig. 1.2 Hierarchical structure of the problem of selecting a food stall

2. *Criteria*: These are the factors used to evaluate the different alternatives. In this case, the criteria are "Profitability," "Popularity," and "Ease of Preparation."
3. *Alternatives*: These are the available options being considered. In this example, the alternatives are "Fried Noodles," "French Fries," "Soft Serve," and "Crepes."

As illustrated in Fig. 1.2, this hierarchical structure visually represents the relationships between the goal, criteria, and alternatives, providing a clear framework for the subsequent analysis.

Pairwise Comparison of the Criteria with Respect to the Goal: In the AHP, determining the relative importance of each criterion is crucial. Pairwise comparisons provide a structured approach to evaluating how each criterion contributes to achieving the decision goal.

The weighting of each evaluation criterion is performed as follows:

1. *Pairwise Comparisons*: Each pair of criteria is compared to assess their relative importance. For example, we might ask, "*Is Profitability more important than Popularity? If so, by how much?*" The relative importance is quantified using a numerical scale, commonly referred to as the fundamental scale of pairwise comparisons (see Table 1.12). This scale provides a standardized and consistent method for expressing judgments, allowing decision-makers to assign values based on experience and reasoning.
2. *Developing a Pairwise Comparison Matrix*: The results of the pairwise comparisons are organized into a matrix, where:
   - Each row and column corresponds to a criterion.
   - The value in each cell represents the relative importance of one criterion compared to another.
   - Reciprocal values are used to maintain consistency (e.g., if Criterion A is twice as important as Criterion B, then the reverse comparison value is 1/2).
3. *Weight Calculation*: The values in the pairwise comparison matrix are used to derive weights for each criterion. These weights indicate the relative importance of each criterion in the overall decision-making process. The weights are normalized to sum to 1, ensuring a proper distribution of total importance across all

Table 1.12 Fundamental scale of pairwise comparisons

Intensity of importance	Definition	Explanation
1	Equal importance	Both elements contribute equally to the objective
3	Moderate importance	Experience or judgment slightly favors one element over the other
5	Strong importance	Experience or judgment strongly favors one element over the other
7	Very strong importance	One element is strongly favored, and its dominance is demonstrated in practice
9	Extreme importance	Evidence favors one element to the highest possible degree
2, 4, 6, 8	Intermediate values	Used when a compromise is needed between two judgments

criteria. The geometric mean method—the most commonly used approach—is employed for this selection process.

Returning to our problem, assume the following relative importance relationships:

- Profitability is moderately more important than Popularity.
- Profitability is strongly more important than Ease of Preparation.
- Popularity is moderately more important than Ease of Preparation.

Based on these qualitative judgments, we can construct the pairwise comparison matrix for the criteria, as shown in Table 1.13.

The features of the pairwise comparison matrix are:

- The diagonal elements are all 1, as each criterion is equally important compared to itself.
- The lower triangle of the matrix contains the reciprocals of the corresponding values in the upper triangle, ensuring consistency in pairwise comparisons.

The geometric mean for each row is calculated as follows:

- Profitability: $(1 \times 3 \times 5)^{\frac{1}{3}} = 2.466$
- Popularity: $\left(\frac{1}{3} \times 1 \times 3\right)^{\frac{1}{3}} = 1.000$
- Ease of Preparation: $\left(\frac{1}{5} \times \frac{1}{3} \times 1\right)^{\frac{1}{3}} = 0.405$

Table 1.13 Pairwise comparisons and weights of criteria

Criterion	Profitability	Popularity	Ease of preparation	Geometric mean	Weight
Profitability	1	3	5	2.466	0.637
Popularity	1/3	1	3	1.000	0.258
Ease of preparation	1/5	1/3	1	0.405	0.105

The weights are derived by normalizing the geometric mean of each criterion relative to the total geometric mean sum:

- Profitability: $\frac{2.466}{2.466+1.000+0.405} = \frac{2.466}{3.871} = 0.637$
- Popularity: $\frac{1.000}{2.466+1.000+0.405} = \frac{1.000}{3.871} = 0.258$
- Ease of Preparation: $\frac{0.405}{2.466+1.000+0.405} = \frac{0.405}{3.871} = 0.105$

These calculated weights represent the relative importance of each criterion based on the pairwise comparisons, which are shown in the last column of Table 1.13.

Pairwise Comparison of the Alternatives with Respect to Each Criterion: We conduct pairwise comparisons to determine the relative importance of each alternative within each criterion. Similar to the process used for comparing criteria, we construct pairwise comparison matrices for each criterion. See Tables 1.14, 1.15, and 1.16.

For each criterion, the weights of the alternatives are calculated using the same process described earlier:

Table 1.14 Pairwise comparison of alternatives for Profitability

Alternative	Fried Noodles	French Fries	Soft Serve	Crepes	Geometric mean	Weight
Fried Noodles	1	5	7	3	3.201	0.566
French Fries	1/5	1	1/3	1/5	0.340	0.060
Soft Serve	1/7	3	1	1/3	0.615	0.109
Crepes	1/3	5	3	1	1.495	0.265

Table 1.15 Pairwise comparison of alternatives for Popularity

Alternative	Fried Noodles	French Fries	Soft Serve	Crepes	Geometric mean	Weight
Fried Noodles	1	5	1/3	1	1.136	0.233
French Fries	1/5	1	1/3	1/5	0.340	0.070
Soft Serve	3	3	1	1/5	1.158	0.238
Crepes	1	5	5	1	2.236	0.459

Table 1.16 Pairwise comparison of alternatives to Ease of Preparation

Alternative	Fried Noodles	French Fries	Soft Serve	Crepes	Geometric mean	Weight
Fried Noodles	1	1/3	1/7	1/3	0.355	0.063
French Fries	3	1	1/3	5	1.495	0.265
Soft Serve	7	3	1	5	3.201	0.568
Crepes	3	1/5	1/5	1	0.589	0.104

- Computing the geometric mean of each row.
- Normalizing the values so that their sum equals 1.

These weights indicate the relative preference for each alternative within the context of a specific criterion.

Making the Decision: The final step in the AHP is to synthesize the weights of the criteria and the weights of the alternatives to determine an overall ranking. See Table 1.17.

The overall rating for each alternative is calculated by summing the weighted scores across all criteria as follows:

- Fried Noodles: $(0.637 \times 0.566) + (0.258 \times 0.233) + (0.105 \times 0.063) = 0.427$
- French Fries: $(0.637 \times 0.060) + (0.258 \times 0.070) + (0.105 \times 0.265) = 0.084$
- Soft Serve: $(0.637 \times 0.109) + (0.258 \times 0.238) + (0.105 \times 0.568) = 0.190$
- Crepes: $(0.637 \times 0.265) + (0.258 \times 0.459) + (0.105 \times 0.104) = 0.298$

Based on the overall ratings, the ranking of alternatives is as follows:

$$\text{Fried Noodles} > \text{Crepes} > \text{Soft Serve} > \text{French Fries}.$$

This ranking provides a quantitative basis for selecting the most suitable food stall for the school festival.

Consistency of Pairwise Comparisons: In the AHP, ensuring the consistency of pairwise comparisons is crucial for reliable results. Inconsistencies in judgments can undermine the accuracy of the outcomes. Inconsistency occurs when the relative importance assigned to elements in different pairwise comparisons contradicts logical relationships.

For instance, if "A is more important than B" and "B is more important than C," logically, "A should be more important than C." If values assigned by the decision-maker fail to align with this logic, the comparisons are deemed inconsistent, and corrective measures may be needed.

To measure consistency, we first derive the *Consistency Index* (*CI*), where a lower value indicates higher consistency. Let the items for pairwise comparison be $a_1, a_2, \ldots, a_M$, and let the pairwise comparison matrix be represented as:

Table 1.17 Weighted scores of alternatives

Criterion	Weight	Fried Noodles	French Fries	Soft Serve	Crepes
Profitability	0.637	0.566	0.060	0.109	0.265
Popularity	0.258	0.233	0.070	0.238	0.459
Ease of Preparation	0.105	0.063	0.265	0.568	0.104
Overall rating		**0.427**	**0.084**	**0.190**	**0.298**

$$B = \begin{pmatrix} b_{11} & \cdots & b_{1M} \\ \vdots & \ddots & \vdots \\ b_{M1} & \cdots & b_{MM} \end{pmatrix}$$

Here, each element b_{ij} is a positive value representing the importance of a_i relative to a_j, satisfying the reciprocal property:

$$b_{ij} = \frac{1}{b_{ji}}, \quad \text{for all } i, j \in \{1, 2, \ldots, M\}$$

Ideally, a pairwise comparison matrix should be perfectly consistent, meaning that for any three elements i, j, k, the following condition holds:

$$b_{ij} = b_{ik} b_{kj}$$

If this condition holds for all comparisons, the values in the matrix can be expressed as ratios of weights:

$$b_{ij} = \frac{w_i}{w_j}, \quad \text{for all } i, j \in \{1, 2, \ldots, M\}$$

where w_i represents the weight of item a_i.

In the geometric mean method, the weight w_m of a_m is calculated as:

$$w_m = \frac{\left(\prod_{k=1}^{M} b_{mk}\right)^{\frac{1}{M}}}{\sum_{k=1}^{M} \left(\prod_{j=1}^{M} b_{kj}\right)^{\frac{1}{M}}}$$

Based on this, we define the *Consistency Index* (*CI*), which quantifies how much the actual pairwise comparison matrix deviates from a perfectly consistent matrix:

$$CI = \frac{c - M}{M - 1}$$

where

$$c = \frac{1}{M} \sum_{m=1}^{M} \left(\frac{1}{w_m} \sum_{k=1}^{M} w_k b_{mk} \right)$$

Here, c approximates the largest eigenvalue of the pairwise comparison matrix, as we will discuss later in the explanation of the Eigenvector Method. We will also see how the *Consistency Index* (*CI*) quantifies the deviation of the actual pairwise comparison matrix from a perfectly consistent one.

Since the value of *CI* depends on the number of elements M, we introduce the *Random Index* (*RI*) for normalization. The *RI* is the average *CI* of randomly generated pairwise comparison matrices with the same number of elements. It serves as a benchmark for assessing the consistency of an actual matrix.

Table 1.18 presents the values of the *Random Index* (*RI*) for different numbers of elements M.

Thomas Saaty generated 500 random matrices for each number of elements (M) from 1 to 12. In these random matrices:

- All diagonal elements are 1 (as each element is equally important to itself).
- Other elements are randomly chosen from the set $\{1/9, 1/8, \ldots, 1/2, 1, 2, \ldots, 8, 9\}$.
- Elements symmetrical with respect to the diagonal are reciprocally related (e.g., if $b_{ij} = 3$, then $b_{ji} = 1/3$).

The *CI* was calculated for each of these 500 randomly generated matrices. The *RI* for a given number of elements M is the average *CI* of these 500 matrices.

Using *RI*, we introduce a normalized metric called the *Consistency Ratio* (*CR*). It is calculated as:

$$CR = \frac{CI}{RI(M)}$$

where $RI(M)$ is the *Random Index* for a matrix with M elements.

The acceptable level of consistency is considered as follows:

- A *CR* value below 0.1 generally indicates acceptable consistency.
- A *CR* value greater than 0.1 suggests that the pairwise comparisons may be inconsistent, requiring a review and revision of the judgments.

Thomas Saaty determined this threshold empirically based on extensive testing. While the threshold is not absolute, it reflects a balance between allowing some inconsistency and maintaining the integrity of the decision-making process.

Example—Calculating the Consistency Ratio: Let us calculate the Consistency Ratio (*CR*) for the pairwise comparisons for Profitability. To improve readability, we introduce symbols in Table 1.14, resulting in Table 1.19.

Since the number of alternatives M is 4, we compute:

$$c = \frac{1}{4}\sum_{m=1}^{4}\left(\frac{1}{w_m}\sum_{k=1}^{4} w_k b_{mk}\right) = 4.235$$

Table 1.18 Random Index (*RI*)

M	1	2	3	4	5	6	7	8	9	10	11	12
RI	0.00	0.00	0.58	0.90	1.12	1.24	1.32	1.41	1.45	1.49	1.51	1.53

Table 1.19 Pairwise comparison matrix of alternatives for Profitability with symbols

Alternative	Fried Noodles	French Fries	Soft Serve	Crepes	Geometric mean	Weight
Fried Noodles	$b_{11} = 1$	$b_{12} = 5$	$b_{13} = 7$	$b_{14} = 3$	3.201	$w_1 = 0.566$
French Fries	$b_{21} = 1/5$	$b_{22} = 1$	$b_{23} = 1/3$	$b_{24} = 1/5$	0.340	$w_2 = 0.060$
Soft Serve	$b_{31} = 1/7$	$b_{32} = 3$	$b_{33} = 1$	$b_{34} = 1/3$	0.615	$w_3 = 0.109$
Crepes	$b_{41} = 1/3$	$b_{42} = 5$	$b_{43} = 3$	$b_{44} = 1$	1.495	$w_4 = 0.265$

The *Consistency Index* (*CI*) is given by:

$$CI = \frac{c - 4}{4 - 1} = 0.078$$

The *Consistency Ratio* (*CR*) is calculated as:

$$CR = \frac{CI}{RI(4)} = \frac{0.078}{0.90} = 0.087 < 0.1$$

Since the calculated *CR* (= 0.087) is less than the acceptable threshold of 0.1, the level of inconsistency in the pairwise comparisons for Profitability is within an acceptable range.

Eigenvector Method: In the above discussion, the weights were calculated using the geometric mean method because it is relatively simple. However, the Eigenvector Method is theoretically fundamental in the AHP for determining the weights of the decision criteria or alternatives and for assessing consistency.

If the pairwise comparison matrix *B* is perfectly consistent, it can be expressed as:

$$B = \begin{pmatrix} w_1/w_1 & \cdots & w_1/w_M \\ \vdots & \ddots & \vdots \\ w_M/w_1 & \cdots & w_M/w_M \end{pmatrix}$$

In this case, we have:

$$\boldsymbol{Bw} = \begin{pmatrix} b_{11} & \cdots & b_{1M} \\ \vdots & \ddots & \vdots \\ b_{M1} & \cdots & b_{MM} \end{pmatrix} \begin{pmatrix} w_1 \\ \vdots \\ w_M \end{pmatrix} = \begin{pmatrix} w_1/w_1 & \cdots & w_1/w_M \\ \vdots & \ddots & \vdots \\ w_M/w_1 & \cdots & w_M/w_M \end{pmatrix} \begin{pmatrix} w_1 \\ \vdots \\ w_M \end{pmatrix} = M\boldsymbol{w}$$

Thus, the weight vector

$$\mathbf{w} = \begin{pmatrix} w_1 \\ \vdots \\ w_M \end{pmatrix}$$

and the number of elements being compared, M, correspond to the eigenvectors and eigenvalues of B.

Since the second and subsequent rows of B are constant multiples of the first row, the rank of B is 1. This means that only one of the eigenvalues is non-zero, while all others are zero. The sum of the eigenvalues equals the sum of the diagonal elements of B, so:

$$\lambda_{max} = M$$

and all other eigenvalues are 0.

Thus, if B is perfectly consistent, the maximum eigenvalue λ_{max} equals M. Otherwise, $\lambda_{max} > M$. Although the proof of this inequality is omitted here, in practice, perfectly consistent pairwise comparisons are rarely achieved. For an inconsistent matrix, λ_{max} is almost always greater than M.

The Eigenvector Method uses the elements of the eigenvector corresponding to the maximum eigenvalue λ_{max} as the weights of the criteria or alternatives. These eigenvectors are normalized so that the sum of their elements equals 1:

$$B\begin{pmatrix} w_1 \\ \vdots \\ w_M \end{pmatrix} = \lambda_{max}\begin{pmatrix} w_1 \\ \vdots \\ w_M \end{pmatrix}, \quad \sum_{m=1}^{M} w_m = 1$$

The difference between λ_{max} and M indicates the consistency of the pairwise comparisons. A larger difference suggests greater inconsistency. The *Consistency Index* (CI) is calculated as:

$$CI = \frac{\lambda_{max} - M}{M - 1}$$

The term $\lambda_{max} - M$ can be considered an index representing the magnitude of the $M - 1$ eigenvalues other than λ_{max}. The above formula gives the average inconsistency per eigenvalue. If there is perfect consistency, this value is zero, but as inconsistency increases, this value becomes larger.

In the geometric mean method, λ_{max} is approximately estimated by c in the formula for calculating the degree of consistency.

Benefits and Challenges of the AHP: The Analytic Hierarchy Process (AHP) is a powerful decision-making tool with numerous advantages. However, it also presents specific challenges that users should be aware of. Tables 1.20 and 1.21 summarize these benefits and challenges.

Table 1.20 Benefits of the AHP

Benefit	Description
Structured approach	AHP provides a structured method for breaking down complex decisions into a hierarchical structure, making them easier to understand and analyze
Quantitative analysis	AHP converts qualitative judgments into quantitative values, enabling numerical comparisons
Consistency checking	AHP includes mechanisms to assess and improve the consistency of pairwise comparisons
Group decision-making	AHP offers a structured framework for collaboration, allowing groups to evaluate and discuss alternatives
Broad applicability	AHP can be applied across diverse fields, such as healthcare, project management, and environmental planning
Transparency and accountability	AHP ensures a clear and well-documented decision-making process, fostering trust among stakeholders
Prioritization of criteria	AHP helps identify and prioritize the most important criteria in decision-making
Evaluation of alternatives	AHP facilitates a comprehensive review of multiple alternatives based on defined criteria
Sensitivity analysis	AHP allows users to assess how changes in input values, such as criteria weights, impact the outcome

1.5 End-of-Chapter Problems

1.5.1 Problem 1.1: Expected Value Paradox

Even though the probability of winning the lottery is very low, people often buy lottery tickets. On the other hand, even when the expected value is very high, people hesitate to bet on Bernoulli's game. Several theories attempt to explain this behavior, but what do you think?

Task: Why do people sometimes not follow expected value norms? Consider this question in the following two cases:

1. Why do people buy lottery tickets when, on average, they are expected to lose money?
2. In the St. Petersburg Paradox (Bernoulli's game), why do people hesitate to spend large amounts of money even though the expected value is infinite?

Table 1.21 Challenges of the AHP

Challenge	Description
Subjectivity	AHP relies on subjective judgments, which may be influenced by biases or varying perceptions
Consistency	AHP requires maintaining logical consistency in pairwise comparisons, a task that becomes increasingly challenging as the complexity of the problem grows
Complexity	AHP's hierarchical structure can be cumbersome and time-consuming for large-scale problems
Data quality	AHP's results depend heavily on the accuracy and completeness of input data
Group decision-making challenges	AHP can lead to challenges in group decision-making due to conflicting opinions and difficulties in reaching consensus
Limited to pairwise comparisons	AHP requires a large number of pairwise comparisons, which grows exponentially for large problems, making the process tedious and sometimes impractical
Difficulty in handling uncertainty	AHP struggles to incorporate uncertainty and risk into the decision-making process
Sensitivity to weighting	AHP is highly sensitive to small changes in criteria weights, which can significantly alter the ranking of alternatives
Potential for manipulation	AHP may be susceptible to intentional or unintentional manipulation, affecting the fairness of outcomes

1.5.2 *Problem 1.2: Hurwicz Criterion*

Section 1.2.2.2 explored the Maximax and Maximin strategies for investment decisions during the school festival using Table 1.6. In the same section, we also examined the Hurwicz criterion, which balances the Maximax and Maximin strategies by incorporating the coefficient of optimism ($0 \leq \alpha \leq 1$).

Task: Using the Hurwicz criterion, determine the optimal investment for different values of α based on Table 1.6.

1.5.3 *Problem 1.3: Minimax Regret*

Minimax Regret is a decision-making criterion that seeks to minimize the disparity between the chosen outcome and the optimal outcome. In other words, it aims to reduce post-decision regret—the sentiment of "I should have chosen differently" upon reviewing the outcome.

Task: Using the minimax regret criterion, find the optimal investment from the combinations listed in Table 1.6.

1.5.4 Problem 1.4: Decision-Making Criteria

We have examined the following decision-making criteria for addressing uncertainty:

1. Expected Value Maximization: Select the option with the highest expected payoff.
2. Maximin (Maximizing the Worst-Case Outcome): Select the option that maximizes the minimum possible outcome, emphasizing a conservative approach.
3. Maximax (Choosing the Best Possible Outcome): Prioritize potential gains by selecting the option with the highest possible return.
4. Minimax Regret: Minimize the potential for regret by selecting the option that reduces the maximum regret.

Task: Under what circumstances would each of these criteria be most appropriate?

1.5.5 Problem 1.5: Prisoner's Dilemma

The Prisoner's Dilemma is a classic scenario in game theory where two individuals, acting in their own self-interest, end up with a worse outcome than if they had cooperated.

Task: Explain why situations like the Prisoner's Dilemma arise, leading individuals to make decisions that result in suboptimal outcomes.

References

Arrow, K. J. (1951). *Individual values and social choices.* Wiley.

Dalkey, N. C. (1969). *The Delphi method: An experimental study of group opinion.* Rand Corporation Research Memorandum, RM-5888-PR. https://doi.org/10.7249/RM5888

Delbecq, A. L., & Van de Ven, A. H. (1971). A group process model for problem identification and program planning. *The Journal of Applied Behavioral Science, 7*(4), 466–492.

Fisher, R., Ury, W., & Patton, B. (1991). *Getting to yes: Negotiating agreement without giving in.* Penguin Books.

Hurwicz, L. (1951). Criteria for decision-making under uncertainty. *Econometrica, 20*(4), 243–257.

Kahneman, D., & Tversky, A. (1979). Prospect theory: An analysis of decision under risk. *Econometrica, 47*(2), 263–292.

Knight, F. H. (1921). *Risk, uncertainty, and profit* (1st ed.). Houghton Mifflin Company.

Kojima, H. (2013). *Mathematical decision techniques.* Asahi Shinsho 442, Asahi Shimbun Publications Inc. (in Japanese)

Lackner, M., & Skowron, P. (2023). *Multi-winner voting with approval preferences.* Springer Briefs in Intelligent Systems. Springer.

Luce, R. D., & Raiffa, H. (1957). *Games and decisions.* Dover Publications.

Nakamori, Y. (2021). *Knowledge technology: Converting data and information into valuable ideas.* Springer.

Nash, J. F. (1950). Equilibrium points in n-person games. *Proceedings of the National Academy of Sciences, 36*(1), 48–51.

Nash, J. F. (1951). Non-cooperative games. *Annals of Mathematics, 54*(4), 289–352.
Osborn, A. (1953). *Applied imagination: Principles and procedures of creative problem solving*. Charles Scribner's Sons.
Saaty, T. L. (1980). *The analytic hierarchy process*. McGraw-Hill.
Savage, L. J. (1954). *The foundations of statistics*. Wiley (1972 Revised version, Dover Publications).
Simon, H. A. (1955). A behavioral model of rational choice. *The Quarterly Journal of Economics, 69*(1), 99–118.
Susskind, L., & Cruikshank, J. (1987). *Breaking the impasse: Consensual approaches to resolving public conflicts*. Basic Books.
von Neumann, J., & Morgenstern, O. (1944). *Theory of games and economic behavior*. Princeton University Press.

Further Reading on Decision-Making Theory

Carrascosa, I. P. (2018). *Large group decision making: Creating decision support approaches at scale*. Springer. This book introduces group decision-making and consensus-building, detailing classical approaches for modeling preferences and making collective decisions.
Dimitrakakis, C., & Ortner, R. (2022). *Decision making under uncertainty and reinforcement learning: Theory and algorithms*. Springer. Focused on recent research, this book delves into decision-making under uncertainty, particularly in reinforcement learning, offering a robust theoretical foundation.
Hirokawa, R. Y., & Poole, M. S. (1996). *Communication and group decision making* (2nd ed.). Sage. This book offers an in-depth overview of group decision-making processes, emphasizing the role of communication in shaping group dynamics and outcomes.
Kochenderfer, M. J., et al. (2015). *Decision making under uncertainty: Theory and application*. The MIT Press. A comprehensive introduction to decision-making under uncertainty, this book approaches the subject from a computational perspective, covering both theory and practical applications.
Osborne, M. J. (2003). *An introduction to game theory*. Oxford University Press. This book provides a concise and accessible introduction to game theory, featuring numerous examples and exercises for clarity and practice.
Roijers, D. M., & Whiteson, S. (2017). *Multi-objective decision making*. Morgan & Claypool. Focused on decision-theoretic planning, this book addresses the challenges of managing multiple objectives, with applications in reinforcement learning algorithms.
Saaty, T. L. (2008). *Decision making for leaders: The analytic hierarchy process for decisions in a complex world* (3rd revised ed.). RWS Publications. This book provides a practical guide to the Analytic Hierarchy Process (AHP), offering valuable insights for leadership and tackling complex decision-making scenarios.
Tadelis, S. (2012). *Game theory: An introduction*. Princeton University Press. A thorough exploration of game theory, this book covers both cooperative and non-cooperative games, blending theoretical depth with practical examples.

Chapter 2
Decision-Making with the Intuitive Mind

Abstract This chapter highlights the essential role of intuition and emotion in decision-making, emphasizing how sensory input and affective processes shape judgments. After exploring how human sensibility influences decision-making, it explains Affective Engineering and Emotional Design, which have gained significant attention in industry for their ability to harness human sensibility to create impactful products and services. The chapter then introduces the possibility model for quantifying emotional and intuitive factors and provides methods for integrating partial evaluations through tools such as ordered weighted averaging operators and non-additive measures.

2.1 Treating Sensibility Rationally

While we strive to make decisions logically, cognitive limitations often prevent us from fully formulating problems with complete information. As a result, we inevitably rely on our sensibility when setting goals and selecting relevant information. This section explores ways to better understand and effectively leverage sensibility in decision-making.

This section begins by defining sensibility and examining methods for its quantification. Next, it introduces a specific quantification method based on sensibility evaluation experiments, which use descriptive terms such as friendly, gorgeous, feminine, and old-fashioned. Additionally, this section highlights recent advancements in Affective Engineering and Emotional Design, two disciplines that have gained significant attention in recent years.

The original version of the chapter has been revised. A correction to this chapter can be found at https:// doi.org/10.1007/978-981-96-6811-3_7

Y. Nakamori, *Sensible Decision-Making*, https://doi.org/10.1007/978-981-96-6811-3_2

2.1.1 Human Sensibility

As we navigate the complexities of decision-making, we encounter the interplay between reason and sensibility. While reason, grounded in logic and analysis, provides a structured framework for problem-solving, sensibility adds a nuanced layer of emotional and intuitive understanding. Together, these elements contribute to the attainment of true wisdom.

Before delving deeper into the meaning of sensibility, it is helpful to explore related concepts:

- Sensitivity: The capacity to be easily affected by emotional or physical stimuli.
- Awareness: Conscious knowledge or perception of something.
- Perception: The process of interpreting sensory information.
- Discernment: The ability to make careful judgments and distinctions.
- Taste: A particular preference or sense of judgment, often in aesthetic matters.
- Appreciation: A recognition of value or an enjoyment of something.
- Intuition: An immediate understanding of something without conscious reasoning.

It is essential to distinguish between sensitivity and sensibility. Sensitivity pertains to perceiving and reacting, whereas sensibility involves reasoning and judgment.

Human sensibility is a multifaceted concept that extends beyond subjective feelings and perceptions. It reflects a complex interplay of various elements, including:

- Perception: The ability to process and interpret sensory information through sight, hearing, smell, taste, and touch. It involves recognizing and evaluating attributes such as color, shape, sound, aroma, flavor, and texture.
- Imagination: The capacity to envision and create ideas or entities that do not yet exist. This ability fosters innovation and supports problem-solving as well as artistic endeavors.
- Aesthetic Sense: The appreciation of beauty, encompassing an understanding of harmony, balance, and order.
- Empathy: The ability to understand and share the emotions of others. Empathy is a fundamental skill for building meaningful relationships and requires putting oneself in another's position.
- Intuition: The ability to make decisions based on feelings and past experiences rather than logical analysis. Often described as a "sixth sense," intuition can enhance creativity and decision-making.
- Culture: The values and perspectives shaped by one's upbringing and social environment. Culture influences both aesthetic sensibilities and ethical values.
- Individuality: The unique traits and characteristics shaped by innate abilities and life experiences. Sensibility is an integral aspect of individuality and varies widely among individuals.

Understanding these components can help us gain a deeper appreciation of the complexity and richness of human sensibility, allowing us to harness it more effectively in decision-making.

2.1.1.1 The Roles of Sensibility in Decision-Making

Sensibility is essential for effective decision-making, complementing logical reasoning by addressing practical, emotional, and long-term considerations. Below are the key roles that sensibility plays in improving decision-making:

1. Bridging the Gap Between Logic and Reality
    - Practicality: Sensibility helps us account for real-world constraints such as time, resources, and potential obstacles.
    - Emotional Intelligence: It enables us to recognize and integrate emotions—such as desires, fears, and hopes—into decision-making.
    - Long-Term Perspective: Sensibility encourages us to think beyond immediate outcomes and consider the lasting effects of our decisions.
2. Enhancing Decision Quality
    - Intuitive Insights: Sensibility provides valuable instinctive insights that may not arise through purely logical analysis.
    - Creative Thinking: It fosters innovative problem-solving by allowing us to explore multiple perspectives and unconventional approaches.
    - Ethical Considerations: Sensibility helps us evaluate the moral implications of our decisions, ensuring alignment with personal and societal values.
3. Human-Centered Decision-Making
    - Empathy: Sensibility enables us to understand and prioritize the needs and perspectives of others.
    - Social Awareness: It helps us assess the broader social impact of our decisions, promoting socially responsible outcomes.

By incorporating sensibility into our decision-making process, we can make more informed, balanced, and human-centered choices. Sensibility enhances decision quality and ensures that our choices reflect both practicality and humanity.

2.1.1.2 Quantifying Human Sensibility

Quantifying human sensibility—particularly subjective experiences such as emotions and feelings—is a complex yet actively researched area. Various methods have been developed, each with its strengths and limitations. Below is an overview of some prominent approaches:

1. Sensibility Evaluation Experiments: In this method, participants interact with products or services and share their impressions through surveys or interviews. Common metrics include:
    - 7-Point Likert Scale (Likert, 1932): Used to measure participants' levels of agreement or satisfaction.
    - Semantic Differential Scale (Osgood et al., 1957): Designed to assess the connotations associated with a concept using bipolar adjective pairs.

    Recent advancements in this field include the application of statistical and machine learning techniques to enhance accuracy. Additionally, online versions of these experiments are being developed to improve accessibility and scalability.
2. Physiological Measurement: This approach quantifies sensibility by measuring physiological responses, such as brainwaves, heart rate, and skin conductance. Recent research in neuroscience and psychology has sought to clarify the relationship between these physiological responses and sensibility. Efforts are also underway to develop more affordable and portable physiological measurement devices, making this method more accessible.
3. Eye Tracking: Eye tracking analyzes participants' gaze patterns to determine where they focus their attention, indirectly quantifying sensibility. Emerging trends in this area include:
    - Integration with Other Methods: Combining eye tracking with other sensibility evaluation techniques for a more comprehensive analysis.
    - Advancements in Precision: The development of higher-precision eye-tracking systems to improve data accuracy.
4. Affective Computing (Emotion AI): Affective computing employs artificial intelligence to analyze data (e.g., images, audio, or text) and infer emotional states. Current research in this field focuses on:
    - Deep Learning: Advances in deep learning technology to improve emotion recognition accuracy.
    - Ethical Considerations: Ongoing debates regarding privacy, bias, and the ethical use of AI in emotion analysis.
5. Other Methods: Additional approaches to quantifying sensibility include:
    - Virtual Reality (VR) and Augmented Reality (AR): These technologies create immersive experiences that can elicit and measure emotional responses.
    - Social Media Analysis: Utilizing data from social platforms to understand user sentiment and sensibility trends.

Applications and Future Directions: Quantifying human sensibility is essential for understanding user experiences, designing resonant products and services, and developing effective marketing strategies. While significant challenges remain, ongoing research continues to drive the development of more accurate and reliable methods.

The ability to objectively analyze sensibility has applications across various fields, including:

Table 2.1 Examples of bipolar scales used in the semantic differential scale

Left word	1	2	3	4	5	6	7	Right word
Soft					✔			Hard
Cool		✔						Warm
Busy			✔					Quiet
Luxury						✔		Simple
Light				✔				Heavy
Dynamic	✔							Static

- Product Development: Designing products that align with user needs and preferences.
- Service Design: Creating customer experiences that foster emotional engagement.
- Marketing: Developing campaigns that effectively connect with target audiences by appealing to their sensibilities.

As research advances, businesses will increasingly benefit from data-driven insights into sensibility, enabling them to develop more meaningful and impactful solutions.

Comparison of Likert and Semantic Differential Scales: Both the Likert and semantic differential scales are widely used in surveys and research, but they serve slightly different purposes:

- *Likert Scale*: This scale measures the level of agreement or disagreement with a statement. Respondents rate their agreement on a predefined scale, typically including *strongly disagree, disagree, neutral, agree,* and *strongly agree.*
- *Semantic Differential Scale:* This scale measures the connotations or perceptions associated with a concept. Respondents evaluate a concept (e.g., a product, service, or brand) on a scale between two bipolar adjectives. For example, respondents might evaluate a product's visual impression on a scale from soft to hard. Table 2.1 provides examples of bipolar scales used in the semantic differential scale, where the numbers in the first row represent the following meanings: 1 = Very soft, 2 = Soft, 3 = A little soft, 4 = Neutral, 5 = A little hard, 6 = Hard, and 7 = Very hard.

2.1.2 *Affective Engineering*

Affective Engineering, initially known as Kansei Engineering, derives its name from the Japanese word *Kansei*, which encompasses concepts such as sensibility and emotion. This discipline emerged in Japan in the late 1970s, when translating *Kansei* into English proved challenging, leading to the retention of its original Japanese term (Nagamachi, 1995, 2002).

Three key factors significantly influenced the development of Affective Engineering:

1. The Need for Differentiation in a Saturated Market: The late 1970s ushered in an era of prosperity following Japan's rapid economic growth. Consumers began prioritizing sensory appeal over mere functionality, intensifying competition among companies striving to differentiate their offerings.
2. The Importance of Human-Centered Design: Traditional product development primarily focused on features and performance, often overlooking user sensibility. As the market matured, companies realized that user satisfaction hinged on creating designs that resonated with consumers on an emotional and sensory level.
3. The Convergence of Psychology, Physiology, and Engineering: Researchers from diverse disciplines—including psychology, physiology, and engineering—collaborated to develop scientific methodologies for understanding human sensibility, laying the foundation for Affective Engineering.

In 1978, Professor Mitsuo Nagamachi of Hiroshima University formally introduced Affective Engineering, defining it as "the discipline of scientifically studying human sensations and emotions and applying that knowledge to the development of products and services."

Since its inception, Affective Engineering has garnered global attention, with research advancing in Europe, Asia, and beyond. Recent developments in the field integrate cutting-edge technologies such as artificial intelligence (AI) and virtual reality (VR).

Affective Engineering is widely applied in fields such as:

- Product Development: Designing products that align with users' emotional and sensory preferences.
- Service Design: Creating user-centered experiences that appeal to human sensibilities.
- Marketing: Developing campaigns that foster emotional connections with target audiences.

By leveraging data-driven insights into user experiences, companies can design products and services that more effectively meet user needs while strengthening their market competitiveness.

2.1.2.1 The Affective Engineering Process

Affective Engineering follows a systematic process to integrate user sensibility into product and service design. The key steps are as follows:

1. *Identifying the Challenge*: Begin by pinpointing specific challenges related to user sensibility. For example, users may find a product visually unappealing or a service difficult to navigate.
2. *User Research*: Conduct research to gain insights into user sensibility. Methods include surveys, interviews, observations, and physiological measurements. The

objective is to understand how users perceive a product or service through their senses.

3. *Extracting Sensibility Elements*: Analyze the collected data to identify sensory factors that shape user experiences. These factors include color, shape, sound, scent, texture, and usability.
4. *Sensibility Evaluation Experiment*: Convert the identified factors into tangible design elements and conduct user testing to assess their effectiveness. Standard evaluation methods include 7-point Likert scales (Likert, 1932), the semantic differential method (Osgood et al., 1957), and paired comparison tests (Thurstone, 1927).
5. *Analysis and Evaluation*: Examine the results of the evaluation experiments to determine whether the design aligns with user sensibility. Analytical techniques such as principal component analysis and qualitative regression analysis are frequently employed to interpret the data.
6. *Refinement*: Based on the analysis, refine the product or service design. This may involve adjusting sensory elements, addressing deficiencies, or conducting additional user testing to optimize the design.
7. *Validation and Assessment*: After the product or service launch, monitor user feedback and market performance to validate the design's success. Continuously assess its alignment with user sensibility and make iterative improvements as needed.

Affective Engineering bridges the gap between science and emotion, enabling companies to create human-centered solutions. As technology advances, this field will continue to expand, allowing for more precise, effective, and emotionally resonant designs.

2.1.2.2 Success Stories of Affective Engineering

Affective Engineering has been successfully applied across various industries, leading to innovative products that resonate with user sensibilities. Below are some notable examples:

- Easy-to-Drink Beer Cans (*Asahi Breweries, Ltd.*)
- LED Ceiling Light (*Panasonic Corporation*)
- Vehicle Development (*Nissan Motor Co., Ltd.*)
- Air-in Shower (*TOTO Ltd.*)

The following provides brief explanations of these innovations:

Easy-to-Drink Beer Cans (***Asahi Breweries, Ltd.***): In 2004, Asahi Breweries, in collaboration with Waseda University, enhanced the drinking experience of beer cans. This development thoroughly pursued the "*ease of drinking*" and "*ease of pouring*" for customers by introducing the concept of Affective Engineering.

To bring this concept to life, the team focused on several innovations:

- Redesigned Spout: A wider, more circular drinking spout was developed, increasing the pouring flow rate by approximately 20%.
- Sensory Evaluation Model: A comprehensive model was established to link sensory perceptions (e.g., "*inflex feeling,*" "*spill feeling*") with measurable physical properties such as "*pouring sound*" and "*drinking mouth area.*"
- Optimized Design: Using the model, the team identified the *drinking mouth area* as the most critical factor in enhancing the overall drinking experience. They tested nine different spout designs through consumer surveys and ultimately selected the optimal design.

Through these efforts, Asahi Breweries successfully developed a beer can that provides a more enjoyable drinking experience, earning positive consumer feedback and enhancing brand appeal.

LED Ceiling Light (***Panasonic Corporation***): In January 2013, Panasonic launched four new LED ceiling lights under its EVERLEDS brand. These lights incorporate Affective Engineering principles to enhance mood and relaxation.

The enhancement was achieved through groundbreaking features, such as:

- Advanced Lighting Modes: Panasonic expanded features such as dimming, color adjustment, and light distribution control by introducing new modes designed to evoke specific emotional states, such as relaxation or focus.
- Sensibility Technology: Using psychological and physiological analysis, Panasonic scientifically assessed how various lighting conditions affect human well-being. This data-driven approach enabled them to optimize lighting settings for maximum comfort and relaxation.

By merging Affective Engineering with cutting-edge lighting technologies, Panasonic has delivered an exceptional user experience, positively influencing mood, comfort, and cognitive performance.

Vehicle Development (***Nissan Motor Co., Ltd.***): Nissan Motor applies Affective Engineering to enhance various aspects of vehicle design, including exterior aesthetics, interior materials, and the user interface (Miki, 2006). This innovative approach integrates sensory evaluation and human behavior analysis, going beyond traditional ergonomic considerations.

In this context, Nissan incorporated creative design solutions, including:

- Interior Design:
  a. Material Selection: Nissan carefully selects materials that are pleasant to the touch and evoke specific emotional responses. Leather is often chosen for its luxurious texture and tactile appeal.
  b. User Interface Design: Switches, levers, and other controls are designed not only for functionality but also for aesthetics. Affective Engineering ensures that these elements are intuitive and provide a satisfying tactile experience.
- Human-Machine Interface:

a. Usability Engineering: Nissan integrates traditional usability engineering techniques with Affective Engineering to create user-friendly interfaces. The company analyzes user behavior and preferences to optimize features such as navigation and infotainment systems.
b. Sensory Evaluation: Nissan assesses the sensory qualities of materials and interfaces to ensure that its vehicles are not only high performing but also emotionally engaging for both drivers and passengers.

By embedding Affective Engineering into its vehicle development process, Nissan designs vehicles that excel in functionality while offering strong sensory appeal and emotional resonance.

Air-in Shower (***TOTO Ltd.***): TOTO Ltd. has redefined the shower experience with its Air-in Shower technology, achieving both comfort and water conservation. TOTO has successfully quantified subjective user experiences by applying Affective Engineering principles, leading to innovative shower solutions (e.g., Okamoto et al., 2012).

This success was made possible through key advancements such as:

- Water-Saving Technology: TOTO's research into water flow manipulation has resulted in showerheads that provide a luxurious shower experience while significantly reducing water consumption.
- Sensory Evaluation: To optimize shower comfort, TOTO developed a detailed evaluation process:

a. Identifying Key Sensory Attributes: A six-item scale was designed to capture essential sensory aspects of the shower experience, including strength, softness, smoothness, weakness, volume, and warmth.
b. Multivariate Analysis: These six attributes were condensed into two primary dimensions: "*sense of irritation*" and "s*ense of volume*." This refinement allowed for a more focused and efficient evaluation process.

TOTO's air-in shower combines cutting-edge technology with a profound understanding of human sensory perception to deliver a refreshing, comfortable, and environmentally friendly bathing experience.

2.1.2.3 Recent Trends in Affective Engineering

Affective Engineering is rapidly evolving, integrating new technologies and methodologies to enhance design and user experience. Key trends shaping its development include:

- AI-Powered Insights: Machine learning techniques analyze user data—such as facial expressions, brainwaves, and social media activity—to gain a deeper understanding of preferences and emotions.

- Immersive VR and AR: Virtual and augmented reality technologies are creating highly realistic testing environments, offering more significant insights into sensory responses and enabling personalized user experiences.
- Sustainable and Well-being-Focused Design: Affective Engineering is increasingly being applied to promote healthier lifestyle choices and environmentally conscious behaviors through sensory-driven design.
- Multimodal Approach: Designers are focusing on the interplay of multiple senses—sight, touch, and sound—to create products that are more engaging and emotionally resonant.

As technology continues to advance, Affective Engineering is poised to revolutionize the way products are designed, tested, and experienced.

2.1.2.4 Emotional Design

Emotional Design (Norman, 2005) is a field that explores how design elements evoke specific emotions in users. Its primary goal is to create products and services that are not only functional but also emotionally engaging.

Key focus areas in Emotional Design include:

- Understanding the Emotional Impact of Design: Researchers analyze how elements such as color, shape, texture, sound, and layout affect user emotions.
- User-Centered Design: Surveys, interviews, and user testing provide valuable insights into user needs and preferences, helping designers craft emotionally resonant experiences.
- The Power of Positive Emotions: Positive emotions—such as joy, satisfaction, and surprise—enhance user experience, increase satisfaction, and foster brand loyalty.
- Guiding User Behavior: Strategic design elements can subtly influence users toward desired actions, such as making a purchase or engaging with specific features.
- Cultural Nuances: Designers must account for cultural differences in aesthetics, color symbolism, and design conventions to ensure products resonate with a global audience.
- Technology-Enhanced Design: Advanced tools like biofeedback sensors and AI enable the measurement and personalization of user experiences, allowing for designs with more significant emotional impact.

By understanding and leveraging the emotional aspects of design, companies can develop products and services that not only fulfill functional needs but also build deep, lasting connections with users.

Affective Engineering vs. Emotional Design: While both Affective Engineering and Emotional Design aim to create emotionally engaging products and services, their approaches differ in several key ways:

- Theoretical Foundation:
  a. Affective Engineering focuses on the scientific relationship between sensory perception and emotion, drawing insights from psychology, physiology, and engineering.
  b. Emotional Design takes a broader interdisciplinary approach, incorporating sociology, anthropology, and psychology to understand the emotional impact of design.
- Methodology:
  a. Affective Engineering primarily relies on quantitative methods such as surveys and physiological measurements.
  b. Emotional Design gathers insights using qualitative methods, including interviews and user testing.
- Focus:
  a. Affective Engineering emphasizes the sensory aspects of design, including sight, touch, sound, smell, and taste.
  b. Emotional Design encompasses a broader range of factors, such as cultural context, brand associations, and user narratives.

Complementary Roles: Although these approaches differ, they complement each other effectively. Affective Engineering provides a scientific foundation for understanding sensory experiences, offering precise and measurable insights. Emotional Design, on the other hand, offers a wider perspective that takes into account the emotional and contextual dimensions of design. Together, they contribute to creating functionally superior and emotionally captivating products that deeply resonate with users.

2.2 Multivariate Analysis to Understand Sensibility

Evaluating a group of products using Affective Engineering techniques begins with identifying a set of sensibility words (primarily adjectives) likely to describe the products. The process involves creating numerous pairs of contrasting adjectives, such as soft/hard or traditional/modern, and applying the semantic differential method to collect evaluations from a panel of assessors.

However, this approach may require data screening to ensure result reliability. For instance, evaluators unfamiliar with the products may introduce bias, particularly when assessing descriptors that require specific knowledge, such as "traditional" or "modern."

Once the data has been screened, multivariate analysis is typically employed to explore and uncover relationships between the following:

- Words: Identifying semantic relationships and underlying dimensions among the adjectives.
- Products: Understanding perceived similarities and differences between the evaluated products.
- Words and Products: Determining which adjectives are most strongly associated with each product.

The multivariate analysis methods often used in Affective Engineering include:

- Principal Component Analysis (PCA) or Factor Analysis: These methods reduce data dimensionality by identifying the key underlying factors that explain variance in evaluators' ratings.
- Qualitative Regression Analysis: This technique models the relationship between product attributes and evaluators' ratings on specific adjectives.

In the following sections, we will explore these analytical techniques in detail.

Notation Definitions: For clarity in the analysis, the following symbols are introduced:

- $o_m, m = 1,2, \cdots, M$: Evaluation objects (e.g., ceramic cups)
- $w_n, n = 1,2, \cdots, N$: Evaluation words (sensibility words)
    - If specified as a *bipolar scale* $\rightarrow w_n : \langle w_n^l, w_n^r \rangle$, where w_n^l is the left word and w_n^r is the right word
    - When used as a variable in statistical analysis $\rightarrow w_n = w_n^r$
- $e_k, k = 1,2, \cdots, K$: Evaluators
- $z_{mnk} \in \{1,2,3,4,5,6,7\}$: Evaluation value based on w_n^r (three-way data)
- E_{mn} : The set of evaluators who rated object o_m using the word w_n
- $|E_{mn}|$: The number of evaluators in E_{mn} (equal to K if there are no missing values)
- z_{mn}: The average evaluation data across evaluators

The average evaluation data, z_{mn}, is calculated as:

$$z_{mn} = \frac{1}{|E_{mn}|} \sum_{e_k \in E_{mn}} z_{mnk}, \quad |E_{mn}| > 0$$

In evaluation experiments, some evaluators skip responses when they find it difficult to assess certain attributes. To account for this, the symbol E_{mn} represents the subset of evaluators who provide ratings.

Subsequent multivariate analyses are performed using these average evaluation data.

2.2.1 Principal Component Analysis

Principal Component Analysis (PCA), introduced by Pearson (1901), is a statistical method for identifying a small number of principal components that capture the maximum variance in a dataset while minimizing information loss. The process is explained below, imaging the hypothetical dataset shown in Table 2.2.

2.2.1.1 The Steps of Principal Component Analysis

Step 1—Defining the Synthetic Variable: A synthetic variable y is defined as a weighted linear combination of the evaluation variables:

$$y = a_1 w_1 + a_2 w_2 + \cdots + a_N w_N$$

The coefficients $a_1, a_2, \cdots, a_N$ are determined to maximize the variance of y, subject to the constraint:

$$a_1^2 + a_2^2 + \cdots + a_N^2 = 1$$

This constraint prevents the variance from becoming arbitrarily large.

Once the synthetic variable is defined, the next step involves simplifying the data for analysis.

Step 2—Centering the Data: To simplify calculations, the data for each variable is centered so that it has a mean of zero:

$$x_{mn} = z_{mn} - \overline{z}_n, \quad \overline{z}_n = \frac{1}{M}\sum_{m=1}^{M} z_{mn}$$

where x_{mn} represents the centered data and $\overline{z}_n$ is the mean of the original data for variable w_n.

Table 2.2 Example data table for principal component analysis

Object	w_1 : ⟨Cute, Tasteful⟩ w_1 = Tasteful	w_2 : ⟨Delicate, Smart⟩ w_2 = Smart	$\cdots$	w_N : ⟨Static, Dynamic⟩ w_N = Dynamic
o_1	$z_{11} = 3.24$	$z_{12} = 5.76$	$\cdots$	$z_{1N} = 4.25$
o_2	$z_{21} = 2.83$	$z_{22} = 6.12$	$\cdots$	$z_{2N} = 5.34$
$\vdots$	$\vdots$	$\vdots$	$\ddots$	$\vdots$
o_M	$z_{M1} = 6.33$	$z_{M2} = 1.98$	$\cdots$	$z_{MN} = 6.29$

With the data centered, attention is now directed toward calculating the variance of the synthetic variable.

Step 3—Variance of the Synthetic Variable: Using the centered data $\{x_{mn}\}$, the values of the synthetic variable are given by:

$$y_m = a_1 x_{m1} + a_2 x_{m2} + \cdots + a_N x_{mN}, \quad m = 1,2, \cdots, M$$

The variance $V(y)$ of y is given by:

$$V(y) = \frac{1}{M}\sum_{m=1}^{M} y_m^2 = \frac{1}{M}\sum_{m=1}^{M} (a_1 x_{m1} + a_2 x_{m2} + \cdots + a_N x_{mN})^2$$

Expanding this expression:

$$V(y) = \sum_{i=1}^{N}\sum_{j=1}^{N}\left\{\frac{1}{M}\sum_{m=1}^{M} x_{mi}x_{mj}\right\}a_i a_j = \sum_{i=1}^{N}\sum_{j=1}^{N} s_{ij}a_i a_j$$

where s_{ij} is the covariance between w_i and w_j.

With the variance expressed in terms of the covariance matrix, the focus now shifts to finding the optimal weights.

Step 4—Optimization Problem: The goal is to maximize $V(y)$ under the constraint:

$$a_1^2 + a_2^2 + \cdots + a_N^2 = 1$$

Using the method of Lagrange multipliers, define the function:

$$F(a_1, \cdots, a_N, \lambda) = \sum_{i=1}^{N}\sum_{j=1}^{N} s_{ij}a_i a_j - \lambda\left(\sum_{i=1}^{N} a_i^2 - 1\right)$$

where λ is the Lagrange multiplier.

The optimization leads naturally to solving the eigenvalue problem, which defines the principal components.

Step 5—Eigenvalue Problem: Taking partial derivatives of F with respect to a_k and setting them to zero yields:

$$\frac{1}{2}\frac{\partial F}{\partial a_k} = \sum_{j=1}^{N} s_{kj}a_j - \lambda a_k = 0, \quad k = 1, 2, \ldots, N$$

Here, the symmetry of the covariance matrix is utilized, where $s_{jk} = s_{kj}$. The above equation can be rewritten as:

$$V\boldsymbol{a} = \lambda\boldsymbol{a}$$

where

$$V = \begin{pmatrix} s_{11} & \cdots & s_{1N} \\ \vdots & \ddots & \vdots \\ s_{N1} & \cdots & s_{NN} \end{pmatrix}, \quad \boldsymbol{a} = \begin{pmatrix} a_1 \\ \vdots \\ a_N \end{pmatrix}$$

Here, V is the covariance matrix, $\boldsymbol{a}$ is the eigenvector, and λ is the eigenvalue. The eigenvalues are non-negative real numbers because V is symmetric and positive semi-definite.

The eigenvector corresponding to the largest eigenvalue gives the coefficients that maximize $V(y)$. With the eigenvectors and eigenvalues identified, the principal components can be derived and subsequently interpreted.

Step 6—Interpretation: The first principal component identifies the primary direction in which the data varies the most. Subsequent components capture additional variance while remaining orthogonal to earlier components. These principal components allow for a simplified yet comprehensive view of the dataset.

2.2.1.2 Principal Component Scores

Let λ_1 be the largest eigenvalue, and let $\boldsymbol{a}_1 = (a_{11}, a_{21}, \cdots, a_{N1})^T$ be its corresponding eigenvector. The synthetic variable, whose coefficients are the elements of the eigenvector $\boldsymbol{a}_1$, is called the *first principal component*:

$$\hat{y}_1 = a_{11}w_1 + a_{21}w_2 + \cdots + a_{N1}w_N$$

Similarly, the *second principal component* is:

$$\hat{y}_2 = a_{12}w_1 + a_{22}w_2 + \cdots + a_{N2}w_N$$

If the first and second principal components explain most of the variance in the data, we can represent each object as a point in a two-dimensional plane:

$$\left(\hat{y}_{m1}, \hat{y}_{m2}\right), \quad m = 1,2, \cdots, M$$

where

$$\hat{y}_{m1} = a_{11}x_{m1} + a_{21}x_{m2} + \cdots + a_{N1}x_{mN}$$

$$\hat{y}_{m2} = a_{12}x_{m1} + a_{22}x_{m2} + \cdots + a_{N2}x_{mN}$$

A scatter plot of these scores provides a visual representation of the relationships between objects, making it easier to identify patterns or clusters in the data.

2.2.1.3 Principal Component Loadings

The correlation coefficient between a principal component and a variable measures the variable's contribution to that principal component. For the p-th principal component, the correlation coefficient is given by:

$$r_{np} = \frac{Cov[w_n, \hat{y}_p]}{\sqrt{Var[w_n]Var[\hat{y}_p]}} = \frac{\sqrt{\lambda_p}a_{np}}{\sqrt{Var[w_n]}}$$

Plotting (r_{n1}, r_{n2}) for $n = 1,2, \cdots, N$ in the two-dimensional space spanned by the first two principal components, we can visualize the relative positions of the variables. This representation provides insights into how the variables contribute to the first two components and their relationships.

Note: Some of the concepts mentioned above, such as the derivation of the correlation coefficient, may require formal proofs that are detailed in specialized statistical literature. For now, the key takeaway is to understand the practical implications of PCA within the context of Affective Engineering.

2.2.1.4 Significance of Principal Component Analysis in Affective Engineering

Principal Component Analysis (PCA) is a powerful statistical technique for dimensionality reduction and data exploration. It provides valuable insights into the underlying data structure, particularly in Affective Engineering, where subjective impressions are analyzed to guide design decisions.

Key implications include:

1. *Dimensionality Reduction*: Sensibility studies often collect data on numerous subjective impressions (e.g., "elegant," "sporty," "modern"), which can be overwhelming to analyze. PCA reduces the dataset's dimensionality while retaining most information, allowing researchers to focus on the most significant factors. This simplification helps identify the key drivers of consumer perceptions.
2. *Identifying Key Sensibility Dimensions*: PCA identifies the most important dimensions underlying consumer responses. These dimensions represent factors that explain variability across subjective impressions. For example, PCA may reveal that terms like "elegant" and "sophisticated" are highly correlated and rely on

the same principal component, suggesting they stem from a common underlying factor, such as "aesthetic refinement."

3. *Visualizing the Emotion Space*: PCA enables the visualization of relationships between sensibility terms and product attributes in a low-dimensional space, often 2D or 3D. This visualization allows researchers to observe how different sensibility terms cluster together or relate to specific product designs, providing a clearer picture of how products are perceived.
4. *Guiding Design Decisions*: PCA offers valuable guidance for design decisions by highlighting the most influential factors affecting consumer perceptions. For example, when designing a new smartphone, Affective Engineering might analyze consumer responses to prototypes using adjectives such as "stylish," "modern," "functional," "user-friendly," and "innovative." PCA may reveal two dominant underlying factors: "aesthetic appeal" and "ease of use." These insights can guide design teams to prioritize features that enhance these aspects, ensuring the final product resonates with consumer preferences.

Conclusion: PCA is an essential tool in Affective Engineering. It helps distill complex consumer impressions into actionable insights. By reducing dimensionality, identifying key sensibility dimensions, and visualizing relationships, PCA provides a robust framework for optimizing product designs to align with consumer expectations and desires.

2.2.2 *Qualitative Regression Analysis*

This section introduces a regression model designed to predict a quantitative objective variable using qualitative explanatory variables (Hayashi, 1952). Table 2.3 provides an example dataset used to build such a model.

In this context:

- $o_1, o_2, \cdots, o_M$: Objects being evaluated
- $r_1, r_2, \cdots, r_K$: Items (qualitative variables)
- $s_{k1}, s_{k2}, \cdots, s_{kC_k}$: Categories of the item r_k

Table 2.3 Example data table for qualitative regression analysis

	Item r_1				Item r_2				$\cdots$	Item r_K				
	s_{11}	s_{12}	$\cdots$	s_{1C_1}	s_{21}	s_{22}	$\cdots$	s_{2C_2}	$\cdots$	s_{K1}	s_{K2}	$\cdots$	s_{KC_K}	
o_1	1	0	$\cdots$	0	0	1	$\cdots$	0	$\cdots$	0	0	$\cdots$	1	y_1
o_2	0	0	$\cdots$	1	1	0	$\cdots$	0	$\cdots$	0	1	$\cdots$	0	y_2
o_3	1	0	$\cdots$	0	0	0	$\cdots$	1	$\cdots$	0	0	$\cdots$	0	y_3
$\vdots$	$\vdots$	$\vdots$	$\cdots$	$\vdots$	$\vdots$	$\vdots$	$\cdots$		$\cdots$	$\vdots$	$\vdots$	$\cdots$	$\vdots$	$\vdots$
o_M	0	1	$\cdots$	0	1	0	$\cdots$	0	$\cdots$	1	0	$\cdots$	0	y_M

- $y_1, y_2, \cdots, y_M$: Values of the objective variable (quantitative outcomes)

Table 2.3 illustrates the structure of a qualitative regression dataset. For each object (o_m), categorical data from qualitative variables (r_k) are represented as binary indicators, and the corresponding objective variable (y_m) provides the quantitative outcome.

***Example*:** Coffee Cup Attributes:

- Item 1—Shape: Categories → Round, Hexagon, Square, etc.
- Item 2—Pattern: Categories → Flower, Geometric, Abstract, Solid, etc.
- Item 3—Internal Color: Categories → Vermilion, White, Black, Blue, etc.

If the objective variable is "Stylishness" (measured on a 1–7 scale), the goals of qualitative regression analysis are:

1. Identify specific combinations of categories that significantly contribute to higher "Stylishness" scores.
2. Quantify the impact of each combination on the "Stylishness" rating.
3. Understand the relative importance of each item (e.g., Shape, Pattern, Internal Color) in determining "Stylishness."

2.2.2.1 Identifying Category Scores

Dummy variables are introduced to represent categorical responses:

$$\delta_m(kq) = \begin{cases} 1, & \text{if object } o_m \text{ belongs to category } s_{kq} \text{ of item } r_k \\ 0, & \text{otherwise} \end{cases}$$

Constraint:

$$\sum_{q=1}^{C_k} \delta_m(kq) = 1, \quad \text{for all } m = 1,2, \cdots, M;\ k = 1,2, \cdots, K$$

This constraint ensures that each evaluation object belongs to exactly one category for each item. An "Other" category is typically included to handle outliers or unclassified responses.

The objective variable is modeled as:

$$\hat{y}_m = \sum_{k=1}^{K} \sum_{q=1}^{C_k} \beta_{kq} \delta_m(kq), \quad m = 1,2, \cdots, M$$

where β_{kq} is the *category score* for a specific category of the item.

The category scores are estimated by minimizing the following objective function:

$$Q = \sum_{m=1}^{M} \left(y_m - \sum_{k=1}^{K} \sum_{q=1}^{C_k} \beta_{kq} \delta_m(kq) \right)^2$$

The optimality condition is given by:

$$\left. \frac{\partial Q}{\partial \beta_{ij}} \right|_{\beta_{ij} = \widehat{\beta}_{ij}} = 0, \quad i = 1,2, \cdots, K; j = 1,2, \cdots, C_i$$

This leads to the following normal equations:

$$\sum_{m=1}^{M} \sum_{k=1}^{K} \sum_{q=1}^{C_k} \widehat{\beta}_{kq} \delta_m(kq) \delta_m(ij) = \sum_{m=1}^{M} y_m \delta_m(ij), \quad i = 1,2, \cdots, K; j = 1,2, \cdots, C_i$$

For each object, there is a constraint that the sum of the dummy variables for each item must be 1, so we set the score of the first category of each item to 0 and solve the normal equations to find the category scores.

Once the category scores $\widehat{\beta}_{kq}$ are obtained, they are normalized as follows:

$$\alpha_{kq} = \widehat{\beta}_{kq} - \frac{1}{M} \sum_{q=1}^{C_k} \widehat{\beta}_{kq} p_{kq}$$

where p_{kq} is the number of responses for category s_{kq} of item r_k.

This normalization step adjusts the category scores so that each item's weighted average becomes zero. This ensures comparability across different categories and aligns the analysis with standard regression practices.

Summary: Qualitative regression analysis identifies significant category combinations and their relative impacts on a quantitative objective variable. Estimating and normalizing category scores provides a robust framework for analyzing qualitative predictors in various contexts, such as product design or consumer behavior studies.

2.2.2.2 Significance of Qualitative Regression Analysis in Affective Engineering

Qualitative regression analysis is a cornerstone of Affective Engineering, providing a robust framework to:

- Quantify Subjective Impressions: Affective Engineering integrates human emotions and perceptions ("sensibility") into product design. Qualitative regression quantifies subjective impressions—such as "stylish," "comfortable," and "elegant"—in relation to specific product attributes.

- Identify Key Design Drivers: Qualitative regression analyzes the coefficients of categorical variables (e.g., colors, materials, and shapes) to identify design choices that significantly influence the desired affective responses.
- Optimize Design Decisions and Predict Consumer Responses: The analysis informs product feature decisions by highlighting the most impactful attributes. For example, if a specific color combination consistently conveys a "luxury" feeling, designers can prioritize that combination in prototypes.
- Deepen Consumer Understanding: Qualitative regression uncovers insights into consumer preferences and their interactions with products by linking design features with consumer sensibilities.

Example: When designing a new car model, Affective Engineering employs qualitative regression to analyze how exterior colors (e.g., red, blue, silver) and interior materials (e.g., leather, cloth, wood) influence consumers' perceptions of key attributes such as "sportiness," "luxury," and "comfort." The results guide the selection of optimal color and material combinations to evoke the desired emotional response in target customers.

2.3 Possibility Modeling of Sensibility Distribution

This section describes how a *possibility model* can represent data collected from a sensibility evaluation experiment. Although a possibility model resembles a probability density model, they are fundamentally different. The possibility model is grounded in Zadeh's membership function for fuzzy numbers.

This section also explores how to recommend suitable products based on customer preferences expressed through sensibility-related terms. For this purpose, it is essential to compute the closeness between two fuzzy numbers. A method for calculating this degree of closeness, rooted in fuzzy theory, is introduced.

However, since the original approach involves complex calculations, this section adopts a simplified personal discrete possibility model, enabling individuals to construct it independently without requiring input from a large group. We will also avoid complex calculations of the closeness between fuzzy sets and make it easier to estimate the degree of compatibility of a product with a customer's desire.

2.3.1 Fuzzy Numbers

A fuzzy set A has a vague boundary defined by a membership function $\mu_A(x)$ (Zadeh, 1965). The membership function $\mu_A(x)$ represents the degree to which x belongs to the fuzzy set A, where

$$0 \leq \mu_A(x) \leq 1$$

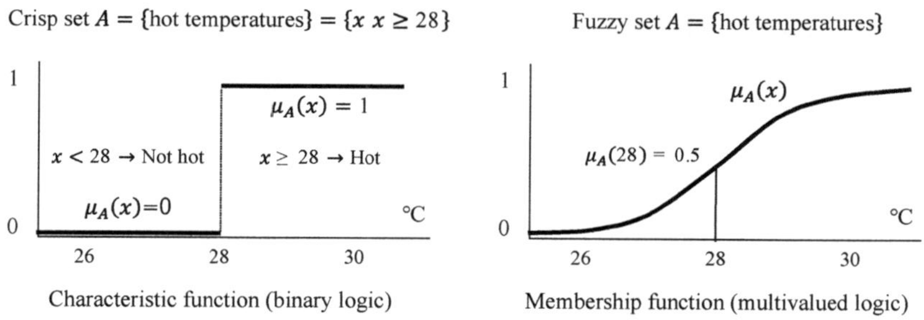

Fig. 2.1 Representation of {hot temperatures} as a crisp set (left) and a fuzzy set (right)

In contrast, for ordinary (crisp) sets, it is defined as:

$$\mu_A(x) = \begin{cases} 1, & \text{if } x \text{ belongs to } A \\ 0, & \text{otherwise} \end{cases}$$

For crisp sets, $\mu_A(x)$ is called the characteristic function.

Example 1 *Fuzzy Set of "Young People"*: Consider the fuzzy set A = {young people}. The membership function assigns *degree of belonging* based on age:

- μ_A(17-year-old) = 0.8
- μ_A(25-year-old) = 1.0
- μ_A(31-year-old) = 0.5

Example 2 *Fuzzy Set of "Hot Temperatures"*: Consider the fuzzy set A = {hot temperatures}. Figure 2.1 illustrates a comparison between:

- A crisp set representation of "hot temperatures" where the boundary is clearly defined.
- A fuzzy set representation, where temperatures gradually transition into "hot" rather than having a strict cutoff.

A *fuzzy number A* is a fuzzy set defined on the real number line that satisfies the following properties:

1. *Normality*: There exists an x such that:

$$\mu_A(x) = 1$$

2. *Convexity*: $\mu_A(x)$ is unimodal, meaning it has a single peak.

A regular number represents a single, precise value, whereas a fuzzy number conveys a range of possible values, each with an associated degree of plausibility.

Example 3 *Fuzzy Set Representing "About 5"*: Consider a fuzzy set:

$$A = \{x \in \mathbb{R} | x \text{ is about } 5\}$$

where $\mathbb{R}$ represents the set of all real numbers. To satisfy the conditions for fuzzy numbers:

1. *Normality*: Assign a membership value of 1 to 5, i.e.,

$$\mu_A(5) = 1.$$

2. *Convexity*: Define the membership function as a triangular shape, as shown on the left side of Fig. 2.2.

In this example, the membership values for 4 and 6 in the fuzzy set "about 5" are:

$$\mu_A(4) = 0.3, \ \mu_A(6) = 0.3$$

2.3.1.1 Possibility Distribution

Zadeh (1978) redefined the membership function as the possibility distribution function, which we refer to in this section as a possibility model. Dubois and Prade (1988) further formalized this concept. Although a rich theoretical framework has been developed around possibility models, this section introduces only the essential aspects relevant to the current discussion.

The possibility distribution function $\mu(x)$ resembles the probability density function $p(x)$, but they are fundamentally different, as previously mentioned.

- A probability density function $p(x)$ integrates to 1 over its entire domain, ensuring total probability.

$$\int_{-\infty}^{+\infty} p(x)dx = 1$$

- A possibility distribution function $\mu(x)$, on the other hand, only guarantees that its maximum value is 1:

$$\max_{x \in X} \mu(x) = 1$$

Example 4 *Possibility Model for "Around 5 O'clock"*: Consider a possibility model for the statement "around 5 o'clock" in the context of meeting someone. The membership function shown on the right side of Fig. 2.2 represents this possibility model.

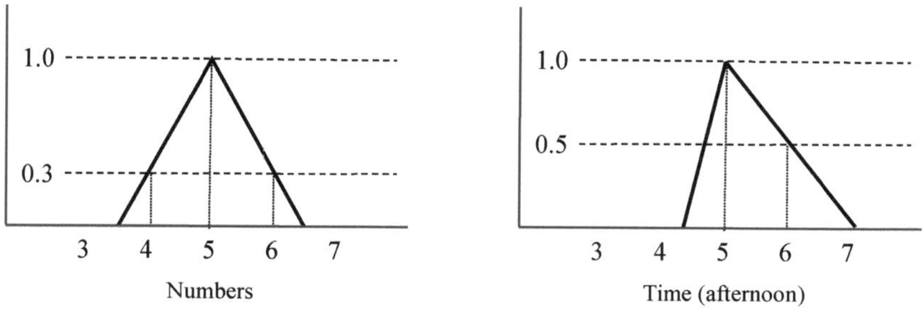

Fig. 2.2 Possibility distribution functions in Example 3 (left) and Example 4 (right)

Even if a person says, "I will meet you at 5 o'clock tomorrow," the possibility function suggests that:

- The possibility of their arriving at exactly 5:00 PM is 1.
- The possibility of their arriving at 6:00 PM is still 0.5.

2.3.1.2 Degree of Closeness Between Fuzzy Numbers

Now, let us examine how to determine the closeness of two fuzzy numbers. For example, suppose a temperature of around 20 degrees Celsius is perceived as comfortable. How would we evaluate the comfort level of a temperature around 25 degrees Celsius?

Let:

- $\mu_A(x)$ be the membership function of the fuzzy set $A = \{\text{comfortable temperatures}\}$.
- $\mu_B(x)$ be the membership function of the fuzzy set $B = \{\text{about 25 degrees Celsius}\}$.

The degree of closeness between "about 25 degrees Celsius" and "comfortable temperature" is calculated as:

$$\max_{x \in X}(\min(\mu_A(x), \mu_B(x))) = \max_{x \in X} \mu_{A \cap B}(x)$$

where X is the set of temperatures within a certain range.

Figure 2.3 illustrates the method of calculating the closeness of two fuzzy numbers.

What does the diagram reveal?

- The comfort level at "exactly 25 degrees Celsius" is relatively low.
- The comfort level at "about 25 degrees Celsius" is slightly higher due to the overlap of fuzzy sets.

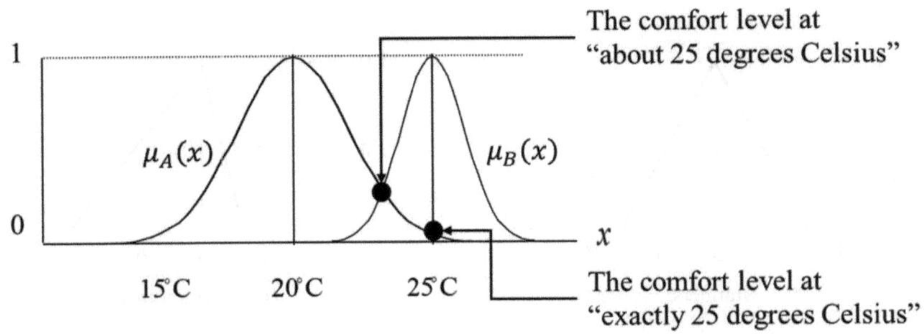

Fig. 2.3 Calculating the closeness of two fuzzy numbers

It is worth noting that the formulas for determining the intersection and union of sets are the same for both fuzzy sets and traditional crisp sets, as given by:

$$\mu_{A\cap B}(x) = \min(\mu_A(x), \mu_B(x))$$

$$\mu_{A\cup B}(x) = \max(\mu_A(x), \mu_B(x))$$

Note: Crisp sets can be considered as a special case of fuzzy sets. In other words, the characteristic function of a crisp set can be viewed as a special form of membership function. Therefore, membership functions and characteristic functions have in common that they express the relationship with the elements of a set.

2.3.2 *Possibility Models*

Building on the earlier preparations, let us address the goal of quantifying sensibility using possibility models. Suppose we have data where multiple evaluators rate various ceramic coffee cups on bipolar adjective scales. The symbols introduced previously are reiterated here for clarity:

- $o_m, m = 1,2,\cdots,M$: Evaluation objects (e.g., ceramic coffee cups)
- $w_n, n = 1,2,\cdots,N$: Evaluation words
  - For bipolar scales, denote $w_n : \langle w_n^l, w_n^r \rangle$, e.g., $w_n : \langle \text{Soft}, \text{Hard} \rangle$
  - When used as a variable in modeling, denote $w_n = w_n^r$, e.g., $w_n = \text{Hard}$
- $e_k, k = 1,2,\cdots,K$: Evaluators
- $z_{mnk} \in \{1,2,3,4,5,6,7\}$: Three-way data

Let z_{mn} denote the average evaluation value used evaluation word $w_n = w_n^r$ for the evaluation object o_m, and let σ_{mn} be the standard deviation. These are used when developing possibility models.

2.3.2.1 Continuous Possibility Models

In this chapter, the continuous *possibility model* $\mu_{mn}(x)$ is specifically defined as follows:

$$\mu_{mn}(x) = \begin{cases} \frac{1}{c\sigma_{mn}}(x - z_{mn} + c\sigma_{mn}), & z_{mn} - c\sigma_{mn} \le x \le z_{mn} \\ -\frac{1}{c\sigma_{mn}}(x - z_{mn} - c\sigma_{mn}), & z_{mn} \le x \le z_{mn} + c\sigma_{mn} \\ 0, & \text{otherwise} \end{cases}$$

where $c > 0$ is a parameter. The value of c is determined based on a subjective judgment informed by sensibility (for example, set $c = 1.5$), but later, it is tuned through actual applications.

Figure 2.4 illustrates this model, which constructs a triangular possibility distribution function using the average and standard deviation of the data.

Possibility Models of Seven Levels: The evaluation values obtained are integers ranging from 1 to 7. However, for example, the evaluation value “3” actually represents “about 3.” Therefore, each level should be treated as a fuzzy number, such as “about 3.”

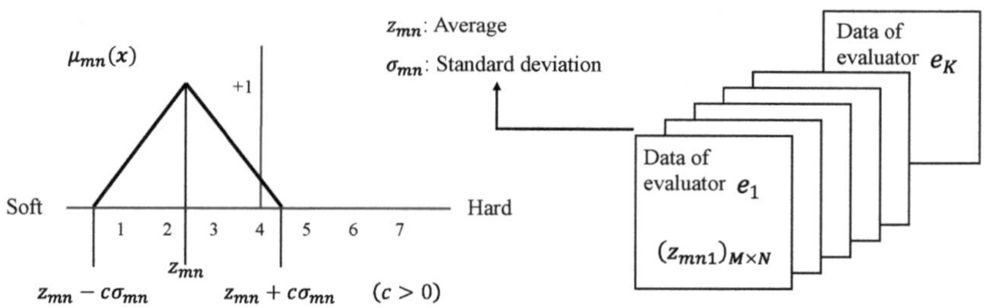

Fig. 2.4 Possibility modeling using data obtained from a sensibility evaluation experiment

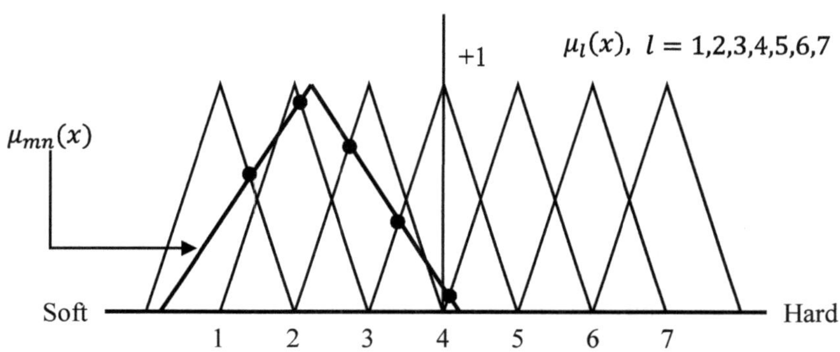

Fig. 2.5 Continuous possibility models

The possibility models for each level l can be defined as follows:

$$\mu_l(x) = \begin{cases} x + 1 - l, & l - 1 \leq x \leq l \\ -x + 1 + l, & l \leq x \leq l + 1 \\ 0, & \text{otherwise} \end{cases}$$

These models are shown in Fig. 2.5, which also presents an example of the possibility model $\mu_{mn}(x)$:

$$\mu_{mn}(x) = \begin{cases} 0.5x - 0.1, & 0.2 \leq x \leq 2.2 \\ -0.5x + 2.1, & 2.2 \leq x \leq 4.2 \\ 0, & \text{otherwise} \end{cases}$$

Although dedicated software could be developed, using continuous possibility models for product recommendations would complicate the calculations. Therefore, we derive discrete possibility models as follows. This enables us to respond immediately when a customer requests to see a soft cup, for example.

2.3.2.2 Discrete Possibility Models

The degree of closeness between the possibility model $\mu_{mn}(x)$ and the possibility model at each level $\mu_l(x)$ is calculated using the following formula:

$$s_{mnl} = \max_x(\min(\mu_{mn}(x), \mu_l(x))), \quad l = 1{,}2{,}3{,}4{,}5{,}6{,}7$$

This formula calculates the maximum degree of overlap between the possibility model $\mu_{mn}(x)$ and each level's possibility model $\mu_l(x)$. The result indicates the closest matching level for the given model, as illustrated in Fig. 2.5.

As a result of the calculation, we obtain:

$$s_{mn1} = 0.6,\ s_{mn2} = 0.93,\ s_{mn3} = 0.73,\ s_{mn4} = 0.4,\ s_{mn5} = 0.067,\ s_{mn6} = s_{mn7} = 0.0$$

Therefore, if w_n is $\langle$Soft, Hard$\rangle$, the following *discrete possibility model* for o_m is obtained:

- Possibility of "Very soft": $s_{mn1} = 0.6$
- Possibility of "Soft": $s_{mn2} = 0.93$
- Possibility of "A little soft": $s_{mn3} = 0.73$
- Possibility of "Neutral": $s_{mn4} = 0.4$
- Possibility of "A little hard": $s_{mn5} = 0.067$
- Possibility of "Hard": $s_{mn6} = 0.0$
- Possibility of "Very hard": $s_{mn7} = 0.0$

So far, we have learned about the "possibility model" based on fuzzy set theory for modeling vague sensibility evaluations. You may ask, "We can accept the center value of the model, but how can we be satisfied with its width (the length of the base of the triangle)?" In practical applications, we refine the model iteratively using new data until we achieve a satisfactory result.

You may also think, "I don't want to spend so much time on this, or I want to choose a product based on my sensibilities, but I don't want to perform complicated calculations." For such individuals, the following subsection will introduce a simple personal discrete possibility model. In this model, the vagueness of the evaluation is kept constant. Additionally, vagueness, such as an evaluation value of "about 3," is not considered.

2.3.2.3 Simple Personal Discrete Possibility Models

Imagine you are evaluating a ceramic coffee cup using a 7-point scale: 1 = Very soft, 2 = Soft, 3 = A little soft, 4 = Neutral, 5 = A little hard, 6 = Hard, and 7 = Very hard.

The possibility at each level is defined as follows:

1. Set the possibility to 1.0 at the level that best matches your perception of the cup on a bipolar scale.
2. For each level away from the best match in either direction, decrease the possibility by 0.2.

For example, if you perceive the coffee cup as "A little soft" (level = 3), the possibilities are assigned as follows:

- Possibility of "Very soft" (level = 1): 0.6
- Possibility of "Soft" (level = 2): 0.8
- Possibility of "A little soft" (level = 3): 1.0
- Possibility of "Neutral" (level = 4): 0.8
- Possibility of "A little hard" (level = 5): 0.6
- Possibility of "Hard" (level = 6): 0.4
- Possibility of "Very hard" (level = 7): 0.2

Similar models can be created for other evaluation axes of the coffee cup. Table 2.4 presents the simple personal discrete possibility models for a coffee cup. Here, $l = 1$, $l = 2$, and so on correspond to levels 1 to 7 on the scale.

Based on Table 2.4, the cup is likely to be perceived as:

- Soft, warm, and very cute.
- Suitable for youth and women.
- Having a high likelihood of being sold successfully.

For example, a cup perceived as soft, warm, and very cute will likely appeal to young women interested in buying it.

Table 2.4 Simple personal discrete possibility models for a coffee cup

Left word	$l = 1$	$l = 2$	$l = 3$	$l = 4$	$l = 5$	$l = 6$	$l = 7$	Right word
Soft	0.6	0.8	1.0	0.8	0.6	0.4	0.2	Hard
Cool	0.0	0.2	0.4	0.6	0.8	1.0	0.8	Warm
Cute	1.0	0.8	0.6	0.4	0.2	0.0	0.0	Tasteful
For youth	0.8	1.0	0.8	0.6	0.4	0.2	0.0	For elderly
For men	0.0	0.0	0.2	0.4	0.6	0.8	1.0	For women
Unsellable	0.0	0.2	0.4	0.6	0.8	1.0	0.8	Sellable

2.3.2.4 Compatibility with Requirements

Suppose a customer requests a cup with the following characteristics:

- Soft ($l = 2$)
- Very warm ($l = 7$)
- A little cute ($l = 3$)
- Suitable for youth ($l = 2$)
- Suitable for women ($l = 6$)

Using Table 2.4, we can determine how well this cup meets the customer's requirements by calculating its degree of compatibility (DC). The results are summarized in Table 2.5.

To apply this method for recommending products that meet multiple sensibility requirements, follow these steps:

- *Step 1*: Identify discrete possibility models for each sensibility word (e.g., $<$ soft, hard $>$), as shown in Table 2.4. If there are multiple candidate products, construct separate tables for each.
- *Step 2*: Using these tables, evaluate each product's compatibility with the customer's requirements and summarize the results in a table like Table 2.5. If there are multiple candidates, this step results in multiple rows of degree of compatibility (DC) values.

In the subsequent section, tables like Table 2.5 will be referred to as "*compatibility tables*," which summarize the degree of compatibility (DC) between product attributes and customer requirements.

Table 2.5 Degrees of compatibility (DC) with customer requirements

	Soft ($l = 2$)	Very warm ($l = 7$)	A little cute ($l = 3$)	For youth ($l = 2$)	For women ($l = 6$)
DC	0.8	0.8	0.6	1.0	0.8

2.4 Decision-Making Through Aggregating Sensibility Evaluations

By incorporating sensibilities into decision-making alongside rational analysis, richer and more nuanced judgments can be achieved. This section explores two methods for making decisions by aggregating sensibility-based evaluation values of products from various perspectives:

- *Aggregation Using Ordered Weighted Averaging (OWA) Operators*: Introduced by Yager (1988), this method provides a flexible approach to combining multiple ratings into a single aggregated value.
- *Aggregation Using Non-Additive Measures*: This method accounts for interactions between individual rating items, enabling a more holistic evaluation.

While this chapter primarily focuses on sensibility evaluation, practical applications allow flexibility in the attributes considered and the data collection methods used.

- *Attributes Considered*: This approach considers not only emotional needs (e.g., cuteness, luxury) but also physical needs (e.g., size, affordability), situational needs (e.g., intended use, suitability for young women), and preference-based needs (e.g., purchase motivation, reputation). By taking these factors into account, product recommendations can be made more comprehensive and better aligned with consumer expectations.
- *Data Collection*: When collecting data, a bipolar scale is not always required, especially when suitable antonyms are difficult to identify. This flexibility ensures a more inclusive and adaptable approach to evaluating and recommending products.

2.4.1 Ordered Weighted Averaging Operators

Imagine you are a store owner selling ceramic tableware. A young customer walks in and says:

> Show me some ceramic coffee cups. They should be somewhat cute, very modern, and not very expensive.

She adds:

> I want a coffee cup that meets my requests as many as possible.

How would you recommend a suitable cup in response to her complex request?

The decision begins by comparing potential coffee cups based on three attributes:

- w_1 : Somewhat cute
- w_2 : Very modern
- w_3 : Not very expensive

Table 2.6 Comparison of two coffee cups from three perspectives

Object	w_1: Somewhat cute	w_2: Very modern	w_3: Not very expensive
Cup o_1	$s_{11} = 0.5$	$s_{12} = 0.3$	$s_{13} = 0.7$
Cup o_2	$s_{21} = 0.6$	$s_{22} = 0.9$	$s_{23} = 0.1$

Suppose you evaluate two coffee cups (o_1 and o_2) and obtain the scores shown in Table 2.6.

If you were to select based on the overall score (e.g., average score), you would select o_2, albeit by a small margin. However, since the customer's priority is to meet "*as many requests as possible*," a more nuanced decision-making approach is needed.

The recommendation depends on how the customer prioritizes their requests, such as:

1. Meets all the requests.
2. Meets as many requests as possible.
3. Meets at least half of the requests.
4. Meets at least one request.

The concept of Ordered Weighted Averaging (OWA) helps address these scenarios. Let us examine the process using the given example.

2.4.1.1 The Idea of Ordered Weighted Averaging Operators

In decision analysis, each attribute (e.g., "somewhat cute") is assigned an attribute evaluation value (e.g., s_{11}). These values are then sorted in descending order:

- For o_1 :
  - Initial scores: $s_{11} = 0.5, s_{12} = 0.3, s_{13} = 0.7$
  - Sorted: $s'_{11} = 0.7, s'_{12} = 0.5, s'_{13} = 0.3$
- For o_2:
  - Initial scores: $s_{21} = 0.6, s_{22} = 0.9, s_{23} = 0.1$
  - Sorted: $s'_{21} = 0.9, s'_{22} = 0.6, s'_{23} = 0.1$

  The overall evaluation value (E_m) is calculated as:

$$E_m = \mu_1 \times s'_{m1} + \mu_2 \times s'_{m2} + \mu_3 \times s'_{m3}, \quad s'_{m1} \geq s'_{m2} \geq s'_{m3}$$

where $m = 1{,}2$ represents the cups, and μ_1, μ_2, μ_3 are weights.

The weights depend on the type of request:

1. *Meets All the Requests*: Use the minimum score as the overall evaluation value:
   - Weights: $\mu_1 = 0.0, \mu_2 = 0.0, \mu_3 = 1.0$

- Results: $E_1 = 0.3(o_1)$, $E_2 = 0.1(o_2)$
- Selected: o_1 (Maximin strategy—guarantees the highest minimum score)

2. *Meets as Many Requests as Possible*: Assign weights such that lower values are prioritized:

   - Weights: $\mu_1 = 0.0$, $\mu_2 = 1/3$, $\mu_3 = 2/3$
   - Results: $E_1 = 1.1/3(o_1)$, $E_2 = 0.8/3(o_2)$
   - Selected: o_1

3. *Meets at Least Half of the Requests*: Assign weights that focus on the higher values:

   - Weights: $\mu_1 = 2/3$, $\mu_2 = 1/3$, $\mu_3 = 0.0$
   - Results: $E_1 = 1.9/3(o_1)$, $E_2 = 2.4/3(o_2)$
   - Selected: o_2

4. *Meets at Least One Request*: Use the maximum score as the overall evaluation value:

   - Weights: $\mu_1 = 1.0$, $\mu_2 = 0.0$, $\mu_3 = 0.0$
   - Results: $E_1 = 0.7(o_1)$, $E_2 = 0.9(o_2)$
   - Selected: o_2 (Maximax strategy—focuses on maximizing the best possible outcome)

Summary: Cup o_1 is recommended if the goal is to meet all or as many requests as possible, whereas cup o_2 is preferable when satisfying at least half or at least one of the requests.

Since the customer wanted a cup that meets "as many requests as possible," you would recommend o_1. The rationale for choosing specific weights will be explained further below.

2.4.1.2 Definition: Ordered Weighted Averaging Operators

Let $\mathbb{R}$ be the set of all real numbers, and let $\mathbb{R}^n$ represent the n-dimensional real coordinate space, where each element is an ordered n-tuple of real numbers.

An OWA operator of dimension n is a mapping:

$$F : \mathbb{R}^n \to \mathbb{R}$$

This operator has an associated collection of weights $\{\mu_1, \mu_2, \cdots, \mu_n\}$, where.

- Each weight lies in the unit interval [0,1].
- The sum of all weights is equal to one.

The OWA operator is defined as:

$$F(s_1, s_2, \cdots, s_n) = \sum_{j=1}^{n} \mu_j s_{j'}$$

where $s_{j'}$ represents the j-th largest value among $\{s_1, s_2, \cdots, s_n\}$ (attribute evaluation values).

Several well-known OWA operators include:

- Maximum operator:

$$F(s_1, s_2, \cdots, s_n) = \max(s_1, s_2, \cdots, s_n) \quad \text{if } \mu_1 = 1 \text{ and } \mu_j = 0 \text{ for } j \neq 1$$

- Minimum operator:

$$F(s_1, s_2, \cdots, s_n) = \min(s_1, s_2, \cdots, s_n) \quad \text{if } \mu_n = 1 \text{ and } \mu_j = 0 \text{ for } j \neq n$$

- Average operator:

$$F(s_1, s_2, \cdots, s_n) = \frac{1}{n} \sum_{j=1}^{n} s_j \quad \text{if } \mu_j = \frac{1}{n} \text{ for all } j \in \{1,2, \cdots, n\}$$

2.4.1.3 How to Determine Weights

To calculate the overall evaluation value using OWA operators, follow these steps:

1. Reorder the attribute evaluation values in descending order.
2. Determine the OWA operator weights using an appropriate method.
3. Aggregate the reordered values using the OWA weights.

Much research has been conducted on how to determine these weights. Below, we introduce Yager's original method (Yager, 1988).

Linguistic Quantifiers: Lotfi Zadeh, the founder of fuzzy set theory, focused on linguistic quantifiers (Zadeh, 1983). These are words or phrases in natural language that express imprecise amounts or proportions, such as *many*, *most*, *few*, or *almost all*. Zadeh's work applied fuzzy logic to represent the inherent vagueness of these quantifiers, allowing for a more nuanced interpretation through degrees of truth rather than strict true/false values.

Table 2.7 outlines four typical linguistic quantifiers and their corresponding membership functions.

When the number of attributes is J, the weights are calculated as:

$$\mu_j = f\left(\frac{j}{J}\right) - f\left(\frac{j-1}{J}\right), \quad j = 1,2, \cdots, J$$

Table 2.7 Linguistic quantifiers and corresponding membership functions

Linguistic quantifier	Membership function
LQ = all	$f(x) = \begin{cases} 0, \ x \neq 1 \\ 1, \ x = 1 \end{cases}$
LQ = as many as possible	$f(x) = \begin{cases} 0, & 0 \leq x \leq 1/2 \\ 2x - 1, & 1/2 \leq x \leq 1 \end{cases}$
LQ = at least half	$f(x) = \begin{cases} 2x, \ 0 \leq x \leq 1/2 \\ 1, \ 1/2 \leq x \leq 1 \end{cases}$
LQ = at least one	$f(x) = \begin{cases} 0, \ x < 1/J \\ 1, \ x \geq 1/J \end{cases}$ J: the number of attributes

Using the membership functions from Table 2.7, the weights for $J = 3$ are:

1. Meets all the requests:

$$\mu_1 = 0.0, \ \mu_2 = 0.0, \ \mu_3 = 1.0$$

2. Meets as many requests as possible:

$$\mu_1 = 0.0, \ \mu_2 = \frac{1}{3}, \ \mu_3 = \frac{2}{3}$$

3. Meets at least half of the requests:

$$\mu_1 = \frac{2}{3}, \ \mu_2 = \frac{1}{3}, \ \mu_3 = 0.0$$

4. Meets at least one request:

$$\mu_1 = 1.0, \ \mu_2 = 0.0, \ \mu_3 = 0.0$$

These weights correspond exactly to those used in the previous example.

2.4.1.4 Example: Recommending the Most Suitable Cup

We begin by determining the rating values for the attributes. The customer's request remains the same:

> Show me some ceramic coffee cups that are somewhat cute, very modern, and not very expensive.

Fig. 2.6 Five traditional ceramic coffee cups to compare

Let us compare five coffee cups, as shown in Fig. 2.6.

The customer's preferences are as follows:

- "Somewhat cute"
- "Very modern"
- "Not very expensive" (e.g., \$30)

The customer evaluates the cuteness of the five cups based on her sensibilities, while the shop owner determines whether each cup is traditional or modern. Table 2.8 presents the evaluation results.

To calculate the overall evaluation values, the qualitative assessments in Table 2.8 must be quantified.

Let us create a simple discrete possibility model for Cup o_1.

1. Tasteful—Cute Axis
   - As shown in Table 2.8, the customer rated Cup o_1 as "Somewhat tasteful or Neutral."
   - Based on this, we can construct a discrete possibility model along this axis, as illustrated in Fig. 2.7. The possibility model can also be presented in a table similar to Table 2.4. Since the possibility is 1 when the level is 3.5 (between Somewhat Tasteful and Neutral), possibilities at the original seven levels are odd numbers, as shown in Fig. 2.7.
   - Since the customer's requirement is "Somewhat cute," the compatibility of Cup o_1 with this requirement is 0.7.

Table 2.8 Evaluation of five ceramic coffee cups

Object	w_1 : ⟨Tasteful, Cute⟩	w_2 : ⟨Traditional, Modern⟩	w_3 : ⟨Expensive, Cheap⟩
Cup o_1	Somewhat tasteful or Neutral	Modern	\$60
Cup o_2	Cute	Neutral or Somewhat modern	\$55
Cup o_3	Tasteful or Somewhat tasteful	Somewhat modern	\$15
Cup o_4	Very cute	Somewhat traditional or Neutral	\$35
Cup o_5	Tasteful or Somewhat tasteful	Modern or Very modern	\$60

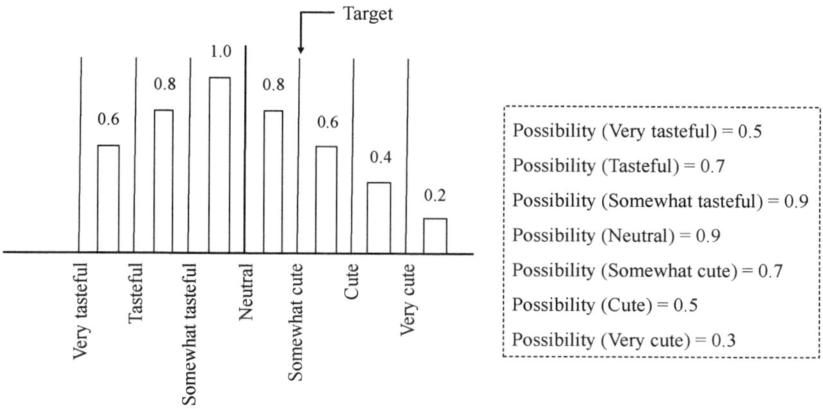

Fig. 2.7 A simple discrete possibility model for Cup o_1 on the axis of tasteful-cute

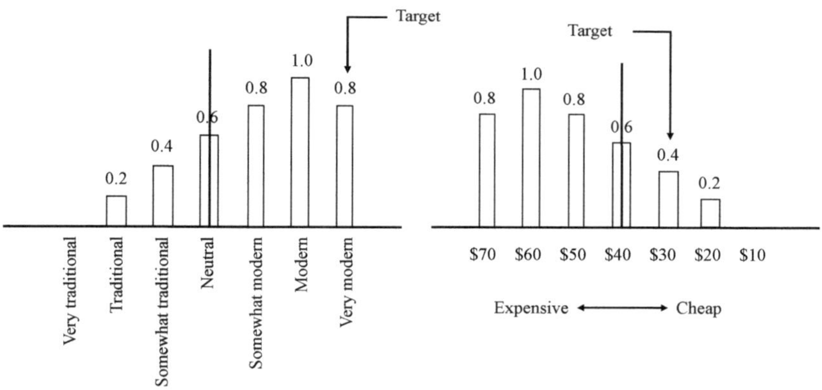

Fig. 2.8 Discrete possibility models for Cup o_1 on the axes of traditional-modern (left) and price (right)

2. Traditional—Modern and Price Axes

- Figure 2.8 presents the discrete possibility model for Cup o_1 on the:
  - Traditional-Modern axis (left)
  - Expensive-Cheap axis (right)
- The customer's requirements are "Very modern" and "Not too expensive" (assume to be \$30).
- Compatibility values for Cup o_1:
  - "Very modern" requirement → 0.8.
  - "Not too expensive" requirement → 0.4.

The compatibility for all five cups is calculated similarly, resulting in Table 2.9.

The compatibility of the five cups can also be determined using a single possibility model. The method described above utilizes membership functions with a height of 1 and a common base, enabling the simultaneous calculation of compatibility for all five cups. Specifically, by constructing a discrete membership function where the request "Somewhat cute" serves as the apex (height 1) and assessing the corresponding heights for the five cups, we can directly obtain the values in the second column of Table 2.9.

Let us now calculate the four selection criteria:

1. Meets all the requests → Find the maximum among the minimums.
2. Meets as many requests as possible → Find the maximum among values given by

$$\frac{2}{3} \times \text{minimum} + \frac{1}{3} \times \text{second minimum}.$$

3. Meets at least half of the requests → Find the maximum among values given by

$$\frac{2}{3} \times \text{maximum} + \frac{1}{3} \times \text{second maximum}.$$

Table 2.9 Compatibility table with requests

Object	w_1: Somewhat cute	w_2: Very modern	w_3: Not very expensive
Cup o_1	$s_{11} = 0.7$	$s_{12} = 0.8$	$s_{13} = 0.4$
Cup o_2	$s_{21} = 0.8$	$s_{22} = 0.5$	$s_{23} = 0.5$
Cup o_3	$s_{31} = 0.5$	$s_{32} = 0.6$	$s_{33} = 0.7$
Cup o_4	$s_{41} = 0.6$	$s_{42} = 0.3$	$s_{43} = 0.9$
Cup o_5	$s_{51} = 0.5$	$s_{52} = 0.9$	$s_{53} = 0.4$

Table 2.10 Calculations for choosing the cup to recommend

Object	w_1	w_2	w_3	Average	All	As many as	At least half	At least one
Cup o_1	0.7	0.8	0.4	**1.9/3 = 0.63**	0.4	1.5/3 = 0.50	2.3/3 = 0.77	0.8
Cup o_2	0.8	0.5	0.5	1.8/3 = 0.60	**0.5**	1.5/3 = 0.50	2.1/3 = 0.70	0.8
Cup o_3	0.5	0.6	0.7	1.8/3 = 0.60	**0.5**	**1.6/3 = 0.53**	2.0/3 = 0.67	0.7
Cup o_4	0.6	0.3	0.9	1.8/3 = 0.60	0.3	1.2/3 = 0.40	**2.4/3 = 0.80**	**0.9**
Cup o_5	0.5	0.9	0.4	1.8/3 = 0.60	0.4	1.3/3 = 0.43	2.3/3 = 0.77	**0.9**

4. Meets at least one request → Find the maximum among the maximums.

   The results are shown in Table 2.10, which shows that:

- Cups o_2 and o_3 satisfy all requests.
- Cup o_3 satisfies as many requests as possible.
- Cup o_4 satisfies at least half of the requests.
- Cups o_4 and o_5 satisfy at least one request.

As stated at the beginning of Sect. 2.4.1, suppose the customer adds the following request:

> I want a coffee cup that meets my requests as many as possible.

In this case, Cup o_3 is recommended. Looking at Fig. 2.6, are you convinced by the result?

However, Table 2.10 shows that by changing the evaluation criteria, any cup has a chance of being recommended, which is quite thought-provoking when we reflect on our own decision-making in life.

2.4.2 Non-Additive Measures

Imagine a new customer arriving and saying,

> Please show me a ceramic cup that feels soft and very warm. I want to give it as a present to my mother, who is in her late 40s.

The customer then adds the following request:

> I would like to include the following three conditions:

- Condition 1: In a single evaluation, “Soft” should be the most important criterion.
- Condition 2: If both “A little for elderly” and “For women” are met, a high score should be given.
- Condition 3: It would be highly desirable if all three criteria—“Soft,” “A little for elderly,” and “For women”—were met.

How would you select a coffee cup that meets these requirements?

This section explores the use of non-additive measures to make decisions in such situations. The above question will be provided as an end-of-chapter problem, so attempt it after learning the techniques.

2.4.2.1 Measures

A measure is defined as follows:

1. Let X be a set. Here, we limit X to a finite set.

2. Let 2^X be the power set of X, which consists of all subsets of X.
3. A measure μ is a function that assigns a non-negative real number to each element of 2^X.

This function maps the size of a set to a non-negative real number and can represent measurements such as length, area, and volume. A measure is also defined as the value produced by this function—that is, the non-negative real number assigned to each set. For example, the measure of length is the size assigned to a line segment. Thus, the term "measure" can refer both to the function that determines the size of a set and to the specific non-negative real number assigned to each set.

A power set is the "set of all subsets" of a given set. The number of elements in the power set is a power of 2. For example, consider a set X with three elements: $\{x, y, z\}$. The number of elements in X is $|X| = 3$. The power set of X consists of all subsets of X, given by:

$$2^X = \{\varnothing, \{x\}, \{y\}, \{z\}, \{x, y\}, \{x, z\}, \{y, z\}, X\}$$

Here, $\varnothing$ represents the empty set, i.e., the set with no elements. The number of elements in 2^X is $2^{|X|}$. In this case, $|X| = 3$, so $2^{|X|} = 8$.

Below are two typical examples of measures:

- *Land area*: Let $A_i (i = 1,2)$ represent 2 pieces of land with no overlapping areas, and let $\mu(A_i)$ represent the land area of A_i. Consider $A_1 \cup A_2$, the total land area of A_1 and A_2. Since A_1 and A_2 are disjoint (i.e., $A_1 \cap A_2 = \varnothing$), the land area of $A_1 \cup A_2$ is the sum of the areas of A_1 and A_2:

$$A_1 \cap A_2 = \emptyset \text{ (Disjoint) } \Rightarrow \mu(A_1 \cup A_2) = \mu(A_1) + \mu(A_2) \text{ (Additivity)}$$

  The total land area is finite.
- *Probability*: Let A_i represent the outcomes of rolling a die, where A_i corresponds to one of the outcomes $\{1, 2, 3, 4, 5, 6\}$. Let $\mu(A_i)$ represent the probability of rolling A_i, which is typically $1/6$. To find the probability of rolling an even number, consider the subsets A_2, A_4, and A_6. These subsets are disjoint, meaning no two of the three events can occur simultaneously. The probability of rolling an even number is:

$$\mu(A_2 \cup A_4 \cup A_6) = \mu(A_2) + \mu(A_4) + \mu(A_6) = \frac{1}{2} \text{ (Additivity)}$$

  The probability of rolling any number from A_1 to A_6 is 1.

  The measure μ has the following properties:

1. *Null Set Property*—The measure of the empty set is zero:

$$\mu(\emptyset) = 0$$

2. *Non-Negativity*—The measure of any subset is non-negative:

$$\mu(A) \geq 0 \quad \text{for any } A \in 2^X$$

3. *Boundedness*—The measure of the entire set is finite. For probability measures,

$$\mu(X) = 1$$

4. *Additivity*—The measure of the union of disjoint sets is the sum of their individual measures:

$$A, B \in 2^X, \; A \cap B = \emptyset \; \Rightarrow \; \mu(A \cup B) = \mu(A) + \mu(B)$$

To address the customer's complex request for a coffee cup, we need to use a measure with relaxed additivity, allowing for flexible evaluation criteria that accommodate overlapping or interacting conditions.

2.4.2.2 Definition: Non-Additive Measures

As described above, the most essential property of a measure is additivity. However, subjective evaluations are often non-additive. Below are two examples that illustrate this concept:

- *Sub-additivity*: In employment exams, expertise in one area is often sufficient to impress the examiners. For instance, both English conversation skills and computer skills are valuable, and excelling in either one can yield a high score. Let x_1 represent English conversation skills, and x_2 represent computer skills. If a candidate possesses both skills, they receive a maximum score of 100. However, possessing only one skill still earns a high score of 90.

  Let μ be a function that maps an individual's ability to a score out of 100:

$$\mu(\{x_1\}) = 90, \; \mu(\{x_2\}) = 90, \; \mu(\{x_1, x_2\}) = 100$$

  Since the following inequality holds:

$$\mu(\{x_1\}) + \mu(\{x_2\}) > \mu(\{x_1, x_2\})$$

this measure is *sub-additive*, indicating that x_1 and x_2 are *substitutable*.

- *Super-additivity*: When purchasing goods, consumers are often influenced by multiple factors simultaneously. Let x_1 represent a product's functionality and x_2 its aesthetics. If both factors are satisfied, the product receives a maximum score of 100. However, if only one factor is satisfied, the product scores only 30.

Let μ be a function that maps a product's attributes (e.g., functionality, aesthetics) to a score out of 100:

$$\mu(\{x_1\}) = 30,\ \ \mu(\{x_2\}) = 30,\ \ \mu(\{x_1, x_2\}) = 100.$$

Since the following inequality holds:

$$\mu(\{x_1\}) + \mu(\{x_2\}) < \mu(\{x_1, x_2\})$$

this measure is *super-additive*, indicating that x_1 and x_2 have a *synergistic* effect, where their combined impact is greater than the sum of their contributions.

Although additivity does not always hold in subjective evaluations, it is reasonable to expect that a more general property—monotonicity—still applies. Monotonicity ensures that adding more elements to a set does not decrease its measure. Formally, a non-additive measure must satisfy the following property:

$$A \subset B \ \Rightarrow\ \mu(A) \leq \mu(B)$$

Since this measure is used as a weighting function in evaluations, it is constrained to values between 0 and 1:

$$\mu : 2^X \rightarrow [0,1]$$

Under these constraints, the non-additive measure is defined as follows:

1. *Null Set Property*—The measure of the empty set is zero:

$$\mu(\emptyset) = 0$$

2. *Non-Negativity*—The measure of any subset is non-negative:

$$\mu(A) \geq 0 \quad \text{for any } A \in 2^X$$

3. *Normality*—The measure of the entire set is finite. By convention, it is defined as follows:

$$\mu(X) = 1$$

4. *Monotonicity*—The measure of a set is at least as large as that of any of its subsets:

$$A, B \in 2^X,\ A \subset B \ \Rightarrow\ \mu(A) \leq \mu(B)$$

This monotonicity property can also be rewritten as:

$$A, B \in 2^X \ \Rightarrow\ \mu(A \cup B) \geq \max(\mu(A), \mu(B))$$

When defining non-additive measures, it is essential to ensure that this inequality holds.

2.4.2.3 Overall Evaluation Using Non-additive Measures

Now, let us consider a method for performing an overall evaluation using non-additive measures. Before introducing non-additive measures, let us first review how additive measures are commonly used in comprehensive assessment.

Consider a set of exam subjects:

$$X = \{w_1, w_2, w_3, w_4\}$$

where

w_1: English, w_2: Mathematics, w_3: Social Studies, w_4: Science.

Let $\mu_1, \mu_2, \mu_3, \mu_4$ be the weights assigned to these subjects. Let us assume an additive measure:

$$\mu : 2^X \to [0,1], \text{ where } \mu_i = \mu(\{w_i\}) \text{ for } i = 1, 2, 3, 4$$

Let the scores of student m in these subjects be $s_{m1}, s_{m2}, s_{m3}, s_{m4}$. The total score of student m is calculated as:

$$E_m = \mu_1 \times s_{m1} + \mu_2 \times s_{m2} + \mu_3 \times s_{m3} + \mu_4 \times s_{m4}$$

Rearranging the summation while assuming the scores satisfy:

$$s_{m3} > s_{m1} > s_{m4} > s_{m2},$$

the overall score can also be expressed as:

$$E_m = \mu_3 \times (s_{m3} - s_{m1}) + (\mu_1 + \mu_3) \times (s_{m1} - s_{m4}) + (\mu_1 + \mu_3 + \mu_4) \times (s_{m4} - s_{m2}) + (\mu_1 + \mu_2 + \mu_3 + \mu_4) \times s_{m2}$$

Since additivity holds, the total score remains the same regardless of how the summation is rearranged.

Figure 2.9 visualizes this process: the left diagram represents the conventional summation of four weighted subject scores, while the right diagram reorders them in descending order before summation, preserving the total score due to additivity.

Now, let us consider aggregating subject scores using a non-additive measure to account for substitutability or synergistic effects among subjects. That is, we replace the additive measure μ with a non-additive measure. However, to ensure meaningful constraints, we assume the following conditions:

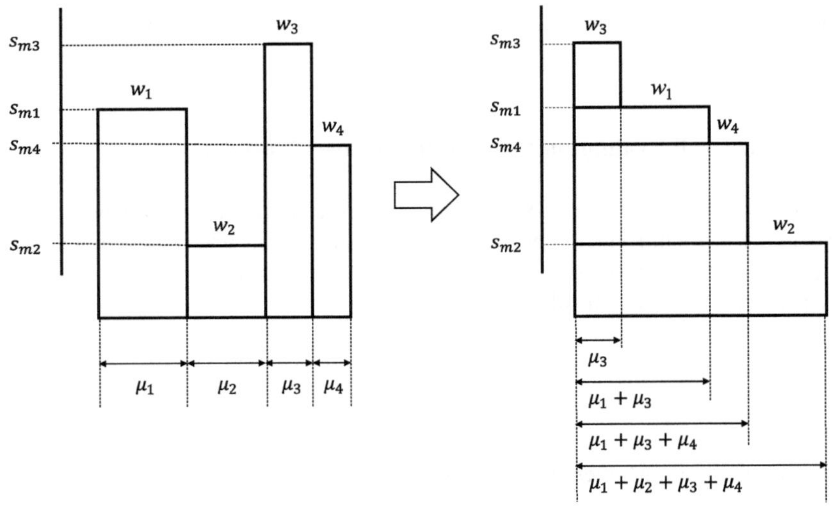

Fig. 2.9 Visualization of overall evaluation using additive measures as weights

$$\mu(\{w_3\}) = \mu_3$$

$$\mu(\{w_1, w_3\}) \neq \mu_1 + \mu_3 \text{ (Non-additive)}$$

$$\mu(\{w_1, w_3, w_4\}) \neq \mu_1 + \mu_3 + \mu_4 \text{ (Non-additive)}$$

$$\mu(\{w_1, w_2, w_3, w_4\}) = \mu_1 + \mu_2 + \mu_3 + \mu_4 = 1$$

These inequalities allow flexibility in defining μ while ensuring that monotonicity is preserved. The values of $\mu(\{w_1, w_3\})$ and $\mu(\{w_1, w_3, w_4\})$ must be selected within a range that maintains the monotonicity condition, ensuring that larger sets have measures at least as great as their subsets.

Using this non-additive measure, the overall score is calculated as:

$$E_m = \mu(\{w_3\}) \times (s_{m3} - s_{m1}) + \mu(\{w_1, w_3\}) \times (s_{m1} - s_{m4}) + \mu(\{w_1, w_3, w_4\}) \times (s_{m4} - s_{m2}) + \mu(\{w_1, w_2, w_3, w_4\}) \times s_{m2}$$

This method of aggregation, which accounts for the interactions between subjects through a non-additive measure, is known as the Choquet integral (Choquet, 1954).

2.4.2.4 Definition: Choquet Integral

Suppose we have a set of evaluation items:

$$X = \{w_1, w_2, \cdots, w_n\}$$

and a corresponding set of evaluation values:

$$S = \{s_1, s_2, \cdots, s_n\}$$

The Choquet integral is used to comprehensively evaluate $s_1, s_2, \cdots, s_n$ by applying a non-additive measure μ that accounts for potential interactions among the evaluation items.

The steps to calculate the Choquet integral are as follows:

1. *Rearrange the Evaluation Values in Descending Order*: Find a permutation σ that sorts the evaluation values $s_1, s_2, \cdots, s_n$ in descending order:

$$s_{\sigma(1)} \geq s_{\sigma(2)} \geq \cdots \geq s_{\sigma(n)}$$

Here, $\{\sigma(1), \sigma(2), \cdots, \sigma(n)\} = \{1,2, \cdots, n\}$. If there are ties in the values, one of them is arbitrarily ranked higher.

2. *Compute the Choquet Integral*: The Choquet integral is a type of *fuzzy integral*, formalized by Murofushi and Sugeno (1991), and is defined as follows:

$$\text{Overall Score} = \sum_{i=1}^{n} \left[\mu\left(\left\{w_{\sigma(1)}, w_{\sigma(2)}, \cdots, w_{\sigma(i)}\right\}\right)\left(s_{\sigma(i)} - s_{\sigma(i+1)}\right)\right]$$

where $\sigma(n+1) = n+1$ and $s_{n+1} = 0$.

In the previous example of test scores, the ranking was given as:

$$s_{m3} > s_{m1} > s_{m4} > s_{m2}$$

According to the Choquet integral definition, the overall score is calculated as follows:

$$E_m = \mu(\{w_3\}) \times (s_{m3} - s_{m1}) + \mu(\{w_3, w_1\}) \times (s_{m1} - s_{m4}) + \mu(\{w_3, w_1, w_4\}) \times (s_{m4} - s_{m2}) + \mu(\{w_3, w_1, w_4, w_2\}) \times s_{m2}$$

Since the elements in a set can be arranged in any order while maintaining the same measure, this can also be written as:

$$E_m = \mu(\{w_3\}) \times (s_{m3} - s_{m1}) + \mu(\{w_1, w_3\}) \times (s_{m1} - s_{m4}) + \mu(\{w_1, w_3, w_4\}) \times (s_{m4} - s_{m2}) + \mu(\{w_1, w_2, w_3, w_4\}) \times s_{m2}$$

2.4.2.5 Example: Selecting an Employee

Imagine four candidates taking an employment exam for a certain company. Table 2.11 shows their exam results. If only one applicant is to be employed, which applicant would you select?

We convert the ratings into a 5-point scale, as shown in Table 2.12. Based on this table, it is difficult to rank the four applicants because the totals, maximums, and minimums are all the same.

To determine the best candidate, we use the Choquet integral, introducing the following conditions:

- Condition 1: The applicant should have strong computer skills (w_3).
- Condition 2: Academic results (w_4) should not outweigh general knowledge (w_1) or interview attitude (w_2).
- Condition 3: A future executive candidate is desired, so a high weight is given if both general knowledge (w_1) and interview attitude (w_2) are strong.
- Condition 4: The highest weight is assigned if computer skills (w_3) are strong in addition to general knowledge (w_1) and interview attitude (w_2).

Considering the above and ensuring monotonicity, the weights are defined as follows:

- Weights for Single Attribute Evaluation

Table 2.11 Evaluations for individual evaluation items

Object	General knowledge (w_1)	Interview attitude (w_2)	Computer skills (w_3)	Academic results (w_4)
Applicant 1 (o_1)	60	Not good	Can do very well	B
Applicant 2 (o_2)	80	Very good	Cannot do much	C
Applicant 3 (o_3)	100	Average	Can do well	D
Applicant 4 (o_4)	40	Good	Average	A

Table 2.12 Ratings using a 5-point scale

Object	General knowledge (w_1)	Interview attitude (w_2)	Computer skills (w_3)	Academic results (w_4)
Applicant 1 (o_1)	$s_{11} = 3$	$s_{12} = 2$	$s_{13} = 5$	$s_{14} = 4$
Applicant 2 (o_2)	$s_{21} = 4$	$s_{22} = 5$	$s_{23} = 2$	$s_{24} = 3$
Applicant 3 (o_3)	$s_{31} = 5$	$s_{32} = 3$	$s_{33} = 4$	$s_{34} = 2$
Applicant 4 (o_4)	$s_{41} = 2$	$s_{42} = 4$	$s_{43} = 3$	$s_{44} = 5$

$$\mu(\{w_1\}) = 0.2,\ \mu(\{w_2\}) = 0.2,\ \mu(\{w_3\}) = 0.5,\ \mu(\{w_4\}) = 0.1$$

- Weights for Pairs of Attributes

$$\mu(\{w_1, w_2\}) = 0.7,\ \mu(\{w_1, w_3\}) = 0.5,\ \mu(\{w_1, w_4\}) = 0.2$$

$$\mu(\{w_2, w_3\}) = 0.5,\ \mu(\{w_2, w_4\}) = 0.2,\ \mu(\{w_3, w_4\}) = 0.5$$

- Weights for Triplets of Attributes

$$\mu(\{w_1, w_2, w_3\}) = 1.0,\ \mu(\{w_1, w_2, w_4\}) = 0.7$$

$$\mu(\{w_1, w_3, w_4\}) = 0.5,\ \mu(\{w_2, w_3, w_4\}) = 0.5$$

- Weights for Empty and Full Sets

$$\mu(\emptyset) = 0.0,\ \mu(\{w_1, w_2, w_3, w_4\}) = 1.0$$

The rationale for defining weights is as follows:

- From Condition 1, set $\mu(\{w_3\}) = 0.5$ (computer skills).
- From Condition 2, set $\mu(\{w_4\}) = 0.1$ (academic results).
- From Condition 3, set $\mu(\{w_1, w_2\}) = 0.7$ (general knowledge and interview attitude).
- From Condition 4, set $\mu(\{w_1, w_2, w_3\}) = 1.0$.
- Other weights are the smallest values that satisfy monotonicity.

The total scores E_1, E_2, E_3, andE_4 for the applicants 1, 2, 3, and 4 are calculated as follows:

- $E_1 = \mu(\{w_3\}) \times (s_{13} - s_{14}) + \mu(\{w_3, w_4\}) \times (s_{14} - s_{11}) + \mu(\{w_1, w_3, w_4\}) \times (s_{11} - s_{12}) + \mu(\{w_1, w_2, w_3, w_4\}) \times s_{12} = 0.5 \times 1 + 0.5 \times 1 + 0.5 \times 1 + 1.0 \times 2 = 3.5$
- $E_2 = \mu(\{w_2\}) \times (s_{22} - s_{21}) + \mu(\{w_1, w_2\}) \times (s_{21} - s_{24}) + \mu(\{w_1, w_2, w_4\}) \times (s_{24} - s_{23}) + \mu(\{w_1, w_2, w_3, w_4\}) \times s_{23} = 0.2 \times 1 + 0.7 \times 1 + 0.7 \times 1 + 1.0 \times 2 = 3.6$
- $E_3 = \mu(\{w_1\}) \times (s_{31} - s_{33}) + \mu(\{w_1, w_3\}) \times (s_{33} - s_{32}) + \mu(\{w_1, w_2, w_3\}) \times (s_{32} - s_{34}) + \mu(\{w_1, w_2, w_3, w_4\}) \times s_{34} = 0.2 \times 1 + 0.5 \times 1 + 1.0 \times 1 + 1.0 \times 2 = 3.7$
- $E_4 = \mu(\{w_4\}) \times (s_{44} - s_{42}) + \mu(\{w_2, w_4\}) \times (s_{42} - s_{43}) + \mu(\{w_2, w_3, w_4\}) \times (s_{43} - s_{41}) + \mu(\{w_1, w_2, w_3, w_4\}) \times s_{41} = 0.1 \times 1 + 0.2 \times 1 + 0.5 \times 1 + 1.0 \times 2 = 2.8$

These results can be interpreted as follows:

- Applicant 1: Strong in Condition 1 but weak in general knowledge and interview attitude, leading to a lower score.
- Applicant 2: Excels in Condition 2, but poor computer skills lower the overall score.

- Applicant 3: Best in Condition 3. A balance of strong computer skills, high scores in general knowledge, and interview attitude results in the highest score.
- Applicant 4: Although the academic performance is good, it is weighted low when evaluated overall, leading to the lowest overall score.

Conclusion: Applicant 3 is the best choice based on the Choquet integral. Unlike simple summation methods, the Choquet integral considers the mutual relationship between evaluation criteria, leading to a decision that better reflects the relative importance of each attribute in a structured manner.

2.5 End-of-Chapter Problems

2.5.1 Problem 2.1: Sensibility in Decision-Making

Sensibility plays a crucial role in decision-making. It allows us to consider nuanced factors beyond pure logic and data, offering intuitive insights that complement rational analysis. By incorporating sensibility-based evaluations, we can recognize patterns, anticipate risks, and make informed judgments based on emotions and past experiences. The integration of logical and intuitive approaches enables us to make more well-rounded and effective choices.

Task: Reflect on the role of sensibility in specific decision-making scenarios.

1. Compare rational and sensibility-oriented approaches when deciding what to eat for lunch.
2. Provide additional examples and discuss how rationality and sensibility complement each other.

2.5.2 Problem 2.2: Affective Engineering

Affective Engineering is the study of designing products and services that elicit specific emotions in users. It focuses on understanding and utilizing human emotions to enhance user experiences. Emotional Design, a closely related concept, examines how emotions shape our interactions with products and environments. It emphasizes addressing both functional and emotional needs, aiming to create products that are not only practical but also enjoyable and satisfying.

Task: To what types of decision-making can Affective Engineering and Emotional Design be applied? Consider the following aspects:

1. Contributions to product development, marketing, and service design.
2. Roles in integrating quantitative and qualitative data, justifying intuitive judgment, and creating new value.

2.5.3 Problem 2.3: Possibility Models

Imagine a customer asks:

> Please show me a ceramic cup that feels soft and very warm. I want to give it as a present to my mother, who is in her late 40s.

As a shop owner, you must recommend one of the five cups shown in Fig. 2.6 (Sect. 2.4.1.4). To do so, you need to assess how well the available cups meet the customer's requirements.

Task: Develop personal discrete possibility models and determine the compatibility of each cup with the customer's requirements by following steps:

1. Create a *personal discrete possibility model* for each of the available cups based on the following four axes:
   - w_1 : ⟨Soft, Hard⟩
   - w_2 : ⟨Cool, Warm⟩
   - w_3 : ⟨For youth, For elderly⟩
   - w_4 : ⟨For men, For women⟩
2. Evaluate *compatibility* by analyzing how well each cup aligns with the customer's requirements along the following axes:
   - "Soft" on the ⟨Soft, Hard⟩ axis
   - "Very warm" on the ⟨Cool, Warm⟩ axis
   - "A little for elderly" on the ⟨For youth, For elderly⟩ axis
   - "For women" on the ⟨For men, For women⟩ axis

2.5.4 Problem 2.4: OWA Operators

Ordered Weighted Averaging (OWA) operators are a family of aggregation functions that enable the flexible combination of multiple values into a single representative value. They are particularly useful when the importance of input values is not fixed but rather depends on their relative ranking. By applying OWA operators, you can determine the most suitable ceramic cup based on the given criteria.

Task: Using the compatibility table obtained in Problem 2.3, apply the OWA operators to identify which of the five cups shown in Fig. 2.6 best satisfies each of the following conditions:

A. Meets all the requests.
B. Meets as many requests as possible.
C. Meets at least half of the requests.
D. Meets at least one request.

2.5.5 Problem 2.5: Choquet Integral

The Choquet integral is a mathematical tool used to aggregate multiple criteria or variables while considering their relative importance and interactions. Unlike traditional weighted averages, which assume independence among criteria, the Choquet integral accounts for dependencies and interactions between them.

Task: Using the compatibility table from Problem 2.3, apply the Choquet integral to recommend a ceramic cup by developing non-additive measures based on the following conditions:

- Condition 1: Among the four criteria, "Soft" should be the most important criterion.
- Condition 2: If both "A little for elderly" and "For women" are satisfied, assign a high score.
- Condition 3: A high score should be given if three criteria—"Soft," "A little for elderly," and "For women"—are satisfied.

References

Choquet, G. (1954). Theory of capacities. *Annales De L'institut Fourier, 5*, 131–295.

Dubois, D., & Prade, H. (1988). *Possibility theory*. Plenum Press.

Hayashi, C. (1952). On the prediction of phenomena for quantitative data and the quantification of qualitative data from the mathematico-statistical point of view. *Annals of the Institute of Statistical Mathematics, 3*, 121–143.

Likert, R. (1932). A technique for the measurement of attitudes. *Archives of Psychology, 22*(140), 1–55.

Miki, Y. (2006). Application of Kansei engineering for vehicle development in NISSAN. *Journal of Japan Society of Kansei Engineering, 6*(4), 35–38. (in Japanese).

Murofushi, T., & Sugeno, M. (1991). A theory of fuzzy measure: Representation, the Choquet integral and null sets. *Journal of Mathematical Analysis and Applications, 159*(2), 532–549.

Nagamachi, M. (1995). Kansei engineering: A new ergonomic consumer-oriented technology for product development. *International Journal of Industrial Ergonomics, 15*, 3–11.

Nagamachi, M. (2002). Kansei engineering as a powerful consumer-oriented technology for product development. *Applied Ergonomics, 33*(3), 289–294.

Norman, D. A. (2005). *Emotional design: Why we love (or hate) everyday things*. Basic Books.

Okamoto, M., Sato, M., & Shodai, Y. (2012). Quantifying comfortable on showering. *Journal of Japan Society of Kansei Engineering, 11*(1), 24–28. (in Japanese).

Osgood, C. E., Suci, G. J., & Tannenbaum, P. H. (1957). *The measurement of meaning*. University of Illinois Press.

Pearson, K. (1901). On lines and planes of closest fit to systems of points in space. *Philosophical Magazine, 2*(11), 559–572.

Thurstone, L. L. (1927). A law of comparative judgment. *Psychological Review, 34*(4), 273–286.

Yager, R. R. (1988). On ordered weighted averaging operators in multicriteria decision making. *IEEE Transactions on Systems, Man and Cybernetics, 18*(1), 183–190.

Zadeh, L. A. (1965). Fuzzy sets. *Information and Control, 8*(3), 338–353.

Zadeh, L. A. (1978). Fuzzy sets as a basis for a theory of possibility. *Fuzzy Sets and Systems, 1*(1), 3–28.

Zadeh, L. A. (1983). A computational approach to fuzzy quantifiers in natural languages. *Computers and Mathematics with Applications, 9*(1), 149–184.

Further Reading on Affective Engineering/Emotional Design

Nagamachi, M. (2010). *Kansei/Affective Engineering*. CRC Press. A foundational work in Kansei Engineering, this book provides detailed explanations of methods and techniques within this key area of affective engineering.

Norman, D. A. (2013). *The design of everyday things: Revised and expanded edition*. Basic Books. Though not exclusively focused on Affective Engineering, this classic text lays a strong foundation in user-centered design principles, highlighting the influence of emotions on user experiences.

Walter, A. (2020). *Designing for emotion: Second edition*. A Book Apart. This book provides a broader perspective on emotional design, particularly in the context of digital experiences, complementing the more specific focus of Kansei Engineering.

Watada, J., et al. (Ed.). (2016). *Industrial applications of affective engineering*. Springer. Offering valuable insights into real-world applications, this book is ideal for students seeking to understand how affective engineering concepts translate into industry practices.

Further Reading on Aggregation Functions

Beliakov, G., Pradera, A., & Galvo, T. (2008). *Aggregation functions: A guide for practitioners* (Studies in Fuzziness and Soft Computing, 221). Springer. While not solely focused on the Choquet integral, this book extensively covers aggregation functions, including fuzzy integrals, and their role in decision-making and other fields.

Grabisch, M., Murofushi, T., & Sugeno, M. (2000). *Fuzzy measures and integrals: Theory and applications* (Studies in Fuzziness and Soft Computing). Physica. This book offers a comprehensive overview of fuzzy measures and integrals, including the Choquet integral, exploring their applications in decision-making, pattern recognition, and image processing.

Yager, R. R., Kacprzyk, J., & Beliakov, G. (Eds.). (2011). *Recent developments in the ordered weighted averaging operators: Theory and practice* (Studies in Fuzziness and Soft Computing, 265). Springer. A detailed exploration of OWA operators, this book covers their theoretical foundations, various extensions, and applications in fields such as decision-making, information fusion, and risk assessment.

Chapter 3
Decision-Making with the Systemic Mind

Abstract This chapter examines systemic decision-making in complex environments. Adopting a systemic mindset allows us to perceive the world as a network of interconnected elements. It explores system models, such as input–output models and causal loop diagrams, to illustrate how relationships and interactions shape outcomes. The discussion also introduces key decision analysis techniques, including decision trees, Bayesian decision theory, and soft systems methodology. Additionally, it highlights the role of big data and artificial intelligence in analyzing intricate systems and enhancing decision-making in dynamic environments.

3.1 Systemic Thinking

Although they have subtle distinctions, "systems thinking," "systematic thinking," and "systemic thinking" are often used interchangeably. Each approach emphasizes a holistic perspective on problems and solutions, recognizing the interconnectedness of various components. Below is a closer look at their meanings:

- *Systems Thinking*: The most commonly used term, referring to an approach that views problems as part of a broader, dynamic system. It emphasizes feedback loops, emergent behavior, and the influence of structure on outcomes. Systems thinking enables us to see the connections.
- *Systematic Thinking*: A structured and organized approach to problem-solving, typically involving a step-by-step methodology. It focuses on processes, organization, and a structured framework. Systematic thinking enables us to follow a process.
- *Systemic Thinking*: A perspective that underscores the interconnectedness of elements within a system, emphasizing relationships and interactions among components. It addresses underlying causes rather than isolated symptoms. Systemic thinking enables us to understand the whole.

Y. Nakamori, *Sensible Decision-Making*, https://doi.org/10.1007/978-981-96-6811-3_3

While the concepts discussed in Chaps. 1 and 2 align more closely with systematic thinking, this chapter focuses on systemic thinking, exploring the intricate web of relationships that underlie complex problems.

3.1.1 *Seeing the World as a System*

The concept of a "system" provides a robust framework for understanding complex entities composed of interconnected parts. By examining systems, we can identify key characteristics that help us analyze and address intricate phenomena:

1. Interconnectedness: Systems emphasize the intricate relationships among their components. For example, an ecosystem is composed of complex interactions between organisms and their environment. Recognizing these connections enables us to predict how changes in one part of the system may affect others.
2. Structure and Organization: Systems operate within underlying structures and frameworks that shape the behavior of their components. Identifying patterns and trends within these structures enhances our ability to analyze and anticipate system behavior.
3. Goal-Oriented Behavior: Many systems operate with specific objectives. For instance, a business seeks to maximize profits and enhance customer satisfaction. By understanding a system's goals, we can assess its performance and pinpoint areas for improvement.
4. Adaptation and Evolution: Systems are dynamic, capable of adapting and evolving over time. Recognizing the systemic nature of challenges enables us to develop more effective and sustainable solutions.

Adopting a systems perspective enhances our ability to understand complex phenomena, make informed decisions, and address challenges holistically and effectively.

3.1.1.1 Viewing Society Through a Systemic Lens

Society can be understood as a collection of interdependent systems, each influencing and shaping human interactions. Among these, economic and social systems play fundamental roles in structuring daily life, shaping opportunities, and determining access to resources. To develop a clearer understanding of these systems, we examine their definitions and key characteristics.

Economic System: The term "economic system" is the framework that shapes economic activity within a society. It emphasizes the interconnectedness of various economic components and the rules, institutions, and mechanisms that influence resource allocation, the production of goods and services, and wealth distribution.

Why emphasize the "system" aspect?

1. Interconnectedness: Economic systems function as complex networks where changes in one area can ripple across the entire system. Recognizing these connections is essential for predicting and managing economic outcomes.
2. Regulatory Framework: Economic systems operate within specific rules, regulations, and norms that shape economic decisions. These structures vary across societies, leading to diverse economic models and outcomes.
3. Systemic Issues: A systems perspective enables us to identify and address the root causes of economic problems rather than merely treating surface-level symptoms.
4. Informing Policymaking: Understanding the economic system helps policymakers design effective policies that enhance financial performance and social welfare while accounting for broader systemic impacts.

By adopting a systems perspective, we gain a deeper understanding of economic complexities, enabling us to develop more effective solutions and promote sustainable development.

Social System: The term "social system" refers to society as a complex network of interconnected individuals and groups. This perspective highlights key aspects that shape human interactions and social structures:

1. Interconnectedness: Individuals and groups are linked through social relationships, influencing each other's behaviors, beliefs, and decisions.
2. Structure and Organization: Society is shaped by institutions such as family, education, religion, and government, which establish roles, expectations, and social order.
3. Culture and Norms: Shared values, beliefs, and customs—collectively known as culture—guide behavior and shape social interactions.
4. Power and Inequality: Power is distributed unevenly, leading to disparities in wealth, status, and opportunities across different social groups.
5. Change and Adaptation: Societies are dynamic and constantly evolving in response to new challenges and opportunities. Social movements often play a crucial role in driving transformation.

Viewing society as a social system allows us to better analyze the forces shaping human behavior, address the root causes of social issues, and explore pathways for positive change.

3.1.1.2 Harnessing Systemic Thinking for Business Success

Systemic thinking provides a powerful framework for managing complex organizational systems. By adopting a holistic perspective, organizations can improve decision-making, enhance performance, and foster innovation. Below are key applications of systemic thinking:

1. Strategic Planning: Analyze the organization and its external environment, including market trends, regulatory changes, and societal factors. Align strategies

with long-term goals and vision. Identify potential disruptions and opportunities and develop contingency plans.
2. Decision-Making: Focus on identifying the root causes of problems rather than merely addressing symptoms. Evaluate how decisions affect different parts of the organization and its external environment. Proactively identify and mitigate internal and external risks.
3. Organizational Design: Create structures that encourage collaboration and information sharing across departments. Empower employees to make decisions and take ownership of their work. Foster a culture of learning and innovation to drive organizational growth.
4. Change Management: Treat change as a systemic process impacting multiple parts of the organization. Engage key stakeholders early to build support for change initiatives. Break down large-scale changes into smaller, manageable steps to facilitate implementation.
5. Customer Relationship Management: Understand the entire customer journey, from initial contact to post-purchase interactions. Gather and utilize customer feedback to refine products and services. Cultivate a customer-centric culture that prioritizes satisfaction and loyalty.
6. Employee Engagement: Take a systemic approach to the employee lifecycle, from recruitment to retirement. Invest in training and development to enhance skills and career growth. Build a supportive work environment to boost morale, productivity, and retention.

Benefits of Systemic Thinking in Corporate Management: By embracing systemic thinking, organizations can achieve the following:

- Improve Decision-Making: Consider the broader context and long-term consequences of decisions.
- Enhance Innovation: Promote creativity, experimentation, and continuous learning.
- Increase Organizational Resilience: Anticipate and adapt to change effectively.
- Strengthen Customer Relationships: Focus on the customer journey and prioritize their needs.
- Boost Employee Engagement and Productivity: Build a positive work culture that supports growth and collaboration.

Systemic thinking is not a one-time solution but an ongoing process that demands continuous learning and adaptation. By integrating this approach, organizations can thrive in today's complex and dynamic business environment.

Walmart—The Retail Powerhouse and Supply Chain Leader: Walmart, founded in 1962 by Sam Walton, is an American multinational corporation that has revolutionized the retail industry with its "Everyday Low Prices" strategy. The company has grown into one of the world's largest and most influential corporations.

A significant factor in Walmart's success is its supply chain mastery. The company uses a systemic approach to oversee a complex network of suppliers, distribution centers, and retail stores, ensuring efficiency and resilience.

The following principles guide Walmart's supply chain strategy:

1. Interconnectedness: Walmart recognizes the interconnected nature of its supply chain. Delays or disruptions at any point—such as at distribution centers—can impact product availability, customer satisfaction, and sales.
2. Feedback Loops: The company employs feedback loops to monitor and improve performance. Metrics such as on-time delivery rates, inventory levels, and customer satisfaction scores help identify areas for improvement.
3. Holistic Perspective: Walmart adopts a system-wide view, focusing on optimizing the entire supply chain rather than isolated components. This approach identifies opportunities for efficiency across the network.
4. Continuous Improvement: Walmart's commitment to continuous improvement drives its efforts to streamline processes, invest in technology, and strengthen supplier relationships, which enhances overall effectiveness.
5. Risk Management: Walmart proactively mitigates supply chain risks, including those related to natural disasters, labor disputes, and geopolitical events. Strategies such as diversifying suppliers and maintaining buffer inventory help reduce potential disruptions.

Key elements of Walmart's systemic approach include:

- Supplier Relationship Management: Walmart collaborates closely with suppliers to ensure a consistent and efficient supply of goods. Emphasis on communication fosters reliability and trust.
- Distribution Center Optimization: Walmart invests heavily in its distribution network, utilizing advanced technology and data analytics to boost efficiency and reduce operational costs.
- Transportation Management: The company's transportation system, including its private fleet and third-party carrier partnerships, ensures the timely delivery of products to stores.
- Inventory Management: Walmart's sophisticated inventory management systems allow it to maintain optimal stock levels, minimize overstocking and stockouts, and reduce costs while improving customer satisfaction.
- Technology Adoption: Walmart leads the retail industry in technology adoption. It leverages innovations such as Radio Frequency Identification (RFID) to track products, advanced analytics tools, and automation to enhance efficiency.

Walmart's systemic approach to supply chain management has been instrumental in its success as a global retail leader. By focusing on interconnectedness, continuous improvement, and technology adoption, Walmart delivers low prices while maintaining high product availability and customer satisfaction. Its innovative practices serve as a benchmark for supply chain excellence worldwide.

3.1.1.3 The Power of Systemic Thinking: A Holistic Perspective

Seeing the world as a system means recognizing its interconnectedness. Every action, decision, and event creates ripples throughout the global network, influencing other parts of the world. Adopting a holistic perspective requires looking at the bigger picture and understanding how different components interact.

Key aspects of systemic thinking are:

- *Interconnectedness*: Recognizing that everything is linked, either directly or indirectly.
- *Holistic View*: Focusing on the entire system rather than isolating individual components.
- *Systems Thinking*: Understanding how systems function, adapt over time, and respond to various inputs.

Why is systemic thinking important?

- Effective Problem-Solving: By analyzing the system as a whole, we can identify root causes and develop solutions that address underlying issues rather than merely treating symptoms.
- Informed Decision-Making: Systemic thinking enables us to anticipate the consequences of our actions and make decisions that benefit the system as a whole.
- Sustainability: A systemic perspective promotes sustainable practices by considering the long-term impacts of our actions on the environment, economy, and society.

Adopting a systemic perspective allows us to approach challenges with a broader understanding, leading to more informed and responsible decision-making. By recognizing the interconnected nature of the world, we can better navigate complexities and create solutions that promote long-term well-being and sustainability.

3.1.2 Systemic Thinking for Decision-Making

Building on the previous discussion, systemic thinking provides a robust framework for decision-making. It considers the interconnectedness of various components by examining problems and solutions within a broader context. This approach emphasizes understanding the entire system rather than focusing solely on individual parts, making it particularly valuable for addressing complex problems involving multiple stakeholders and variables.

3.1.2.1 Key Principles of the Systemic Thinking

The following key principles of systemic thinking are essential for decision-making:

A. *Interconnectedness*: Recognizing that all elements within a system are connected. Decisions made in one area can create ripple effects across other parts of the system, often leading to unintended consequences.
B. *Feedback Loops*: Understanding how actions influence future outcomes through feedback loops. Analyzing these loops helps identify patterns and trends, allowing for adjustments that improve long-term outcomes.
C. *Holism*: Considering the entire system rather than isolating individual parts. A holistic perspective enables a deeper understanding of problems and reveals solutions that benefit the system as a whole.
D. *Emergent Properties*: Acknowledging that systems can exhibit behaviors and characteristics that are not predictable from their components. This principle highlights the importance of viewing the system as an integrated whole rather than a collection of independent parts.

As we have seen, to deal with economic, social, or business systems, it is essential to recognize interconnectedness and feedback loops and take a holistic view. Understanding emergent properties is also crucial for systemic decision-making. Below, we explore these concepts in greater depth.

Interconnectedness—Everything Is Linked: Interconnectedness is a cornerstone of systemic thinking, which recognizes that all elements within a system are linked. Actions in one part of a system can create ripple effects throughout, influencing outcomes in ways that may be unexpected or unintended. Decisions made in isolation often fail to account for these dynamics, leading to unforeseen consequences.

By embracing interconnectedness, decision-makers can:

- Identify Unintended Consequences: Anticipate the potential ripple effects of their decisions on other parts of the system and minimize negative side effects.
- Make Informed Choices: Evaluate the broader implications of their actions to select options that are more likely to achieve desired outcomes.
- Foster Collaboration and Coordination: Work with other stakeholders within the system to align efforts, achieve shared goals, and prevent unintended conflicts.
- Promote Sustainability: Consider the long-term impacts of decisions to ensure the system's continued functionality and health over time.

Understanding interconnectedness is crucial when addressing environmental challenges. Examples include:

- Deforestation: Clearing forests for agriculture or development increases carbon dioxide emissions, contributing to global warming. This triggers altered weather patterns, which, in turn, impact agriculture and ecosystems worldwide.
- Overfishing: The excessive harvesting of certain fish species disrupts marine food chains, leading to declining populations of other species and even the collapse of entire ecosystems.

Interconnectedness underscores the importance of a holistic perspective in decision-making, particularly in complex systems. By considering the relationships between components and the broader context, decision-makers can develop more informed, effective, and sustainable solutions to challenges.

Feedback Loops—Learning from the Past, Shaping the Future: A feedback loop is a mechanism through which a system's output influences its input, creating a cycle of cause and effect. Feedback loops are integral to understanding how systems evolve and respond to changes. There are two main types:

- *Positive Feedback Loops*: These amplify changes, leading to exponential growth or decline. For example, rising temperatures cause more polar ice to melt, exposing oceans and land. These surfaces absorb more sunlight, further increasing temperatures. This self-reinforcing cycle accelerates global warming.
- *Negative Feedback Loops*: These counteract changes, helping maintain stability or equilibrium. For example, rising temperatures cause more evaporation, which in turn leads to cloud formation. These clouds reflect sunlight, cooling the planet. While this helps reduce warming, it does not entirely reverse the overall trend.

Understanding feedback loops is crucial in systemic decision-making because they enable us to:

- Identify Unintended Consequences: Positive feedback loops can amplify changes, sometimes leading to unexpected and undesirable outcomes.
- Predict System Behavior: By analyzing feedback loops, decision-makers can anticipate how a system will respond to inputs and interventions, helping predict long-term impacts.
- Design Effective Interventions: Strategic manipulation of feedback loops can stabilize systems or achieve desired outcomes.

A clear example of feedback loops can be found in the economy:

- Positive Feedback Loop: Rising stock prices encourage more investment, which leads to further market growth.
- Negative Feedback Loop: Raising interest rates discourages borrowing and spending, which slows economic growth.

Understanding these feedback loops helps policymakers design interventions to stabilize the economy, minimize risks, and guide economic growth in a sustainable direction.

Feedback loops are fundamental to understanding how systems operate and evolve. By recognizing and analyzing the dynamics of positive and negative feedback loops, decision-makers can better anticipate system behavior, avoid unintended consequences, and design effective interventions for long-term success.

Holism—Seeing the Big Picture: Holism in systemic decision-making emphasizes considering the entire system rather than focusing solely on individual parts. It recognizes the interconnectedness of all elements within a system and how they influence one another.

Key aspects of a holistic approach include:

- *Considering the Whole System*: Analyze how various components of the system interact and influence one another.
- *Understanding Emergent Properties*: Acknowledge that systems can exhibit behaviors or properties that are not apparent when examining individual components in isolation.
- *Avoiding Reductionism*: Recognize that breaking a complex system down into its simplest parts may fail to capture its complete behavior and dynamics.

By adopting a holistic perspective, decision-makers can:

- Make More Informed Choices: Consider the broader context and interactions within the system to anticipate potential consequences and make decisions aligned with desired outcomes.
- Foster Sustainability: Identify and address systemic issues with long-term impacts, encouraging sustainable practices and solutions.
- Promote Collaboration and Coordination: Understand the interconnected nature of different system components to encourage stakeholder collaboration and align efforts toward shared goals.

A holistic approach to healthcare illustrates the importance of considering the "big picture." Rather than focusing only on individual health concerns, it incorporates:

- Social Determinants of Health: Factors such as income, education, and living conditions that influence health outcomes.
- Access to Healthcare Services: Ensuring equitable availability of medical resources.
- Environmental Conditions: Addressing environmental factors that impact overall well-being.

By recognizing the interconnectedness of mental and physical health, as well as social and environmental factors, a holistic approach enables comprehensive and effective management of health issues.

Holism in systemic decision-making involves acknowledging the complexity and interconnectedness of systems. By seeing the big picture, decision-makers can address challenges comprehensively, foster sustainability, and develop coordinated solutions that benefit the entire system.

Emergent Properties—Beyond the Sum of Their Parts: Emergent properties are characteristics or behaviors that arise from the interactions between a system's components. They are not inherent to any individual component but emerge from their complex interactions. In simpler terms, emergent properties refer to unexpected or surprising outcomes resulting from the interaction of various elements within a system. While these outcomes can be challenging to predict or control, understanding them is essential for effective decision-making.

Traffic congestion in a city is an example of an emergent property. It results from the interactions of individual drivers, road infrastructure, and traffic signals. While it is impossible to predict each driver's exact behavior, the collective actions of all drivers can lead to congestion. This phenomenon emerges from the system as a whole.

By understanding emergent properties, decision-makers can:

- Avoid Reductionism: Prevent the assumption that a system can be fully understood by studying its components in isolation.
- Anticipate Unexpected Outcomes: Prepare for potential surprises or unintended consequences that may arise from the complex interactions within a system.
- Promote Adaptability: Design flexible and resilient systems capable of adapting to new challenges and opportunities.

Emergent properties can help explain market crashes, at least in part. A financial market is a complex system of interacting agents, including investors, traders, and financial institutions. Each agent makes decisions based on available information and individual objectives. However, the collective behavior of all agents can lead to outcomes—such as market crashes—that cannot be predicted by analyzing individual actions alone.

Emergent properties highlight the complexity and interconnectedness of systems. By recognizing the potential for unexpected outcomes, decision-makers can make more informed, adaptable, and resilient choices.

3.1.2.2 Techniques Supporting Systemic Thinking

Table 3.1 provides examples of techniques for applying systemic thinking. By utilizing these techniques and developing the associated skills, you can strengthen your systemic thinking mindset and effectively address various challenges and opportunities.

3.1.2.3 Applying Systemic Thinking

Applying systemic thinking to both research planning and the job searching process can provide a more comprehensive and strategic approach. By recognizing the interconnectedness of various factors and considering the broader context, you can make more informed decisions, enhance problem-solving, and achieve more meaningful outcomes.

Research Planning: Systemic thinking provides a powerful framework for enhancing the effectiveness and impact of research planning. By considering the broader context and the interconnectedness of various factors, researchers can make more informed decisions, identify innovative research questions, and anticipate potential challenges.

Systemic thinking can enhance research planning in several ways:

Table 3.1 Techniques supporting systemic thinking

Category	Technique	Description	References
Visual thinking	Causal loop Diagrams	A visual tool that illustrates how variables influence each other, highlighting feedback loops	Sterman (2000), Forrester (1961)
	Mind maps	A technique for brainstorming and organizing ideas visually	Buzan and Buzan (1996)
	Rich pictures	Hand-drawn diagrams that capture system complexity, including actors, relationships, and structures	Checkland (1981), Checkland and Scholes (1990)
Analytical thinking	System dynamics	A quantitative modeling approach for analyzing complex systems over time, identifying patterns, trends, and the impacts of interventions	Meadows et al. (1972), Morecroft (2007)
	SWOT analysis	A strategic planning tool for identifying strengths, weaknesses, opportunities, and threats	Dess et al. (2005)
	Scenario planning	A method for creating different future scenarios to anticipate challenges and opportunities	Schwartz (1996)
Collaborative tools	Group model building	A collaborative approach to developing a system model	Vennix (1996)
	World café	A structured dialogue process designed to encourage open and honest communication	Brown and Isaacs (2005)
	Appreciative inquiry	A strengths-based approach to organizational change that focuses on what works well	Cooperrider and Whitney (2005)
Systemic thinking frameworks	Soft systems methodology	A structured approach to understanding complex social systems	Checkland (1981), Wilson (1990)
	Viable system model	A framework for analyzing the core functions necessary for system viability	Beer (1979, 1984)

1. Problem Formulation: Instead of focusing on isolated problems, consider the broader system and its underlying causes. Encourage interdisciplinary collaboration to gain diverse perspectives and better understand the issue's complexity.
2. Research Design: Recognize the interconnectedness of variables within the system. Identify positive and negative feedback loops to understand how systems self-regulate. Consider the research's long-term social, economic, and environmental implications.
3. Data Collection and Analysis: Collect data that captures the broader context of the research problem. Combine qualitative and quantitative methods to gain a

comprehensive understanding. Utilize tools such as system dynamics and causal loop diagrams to analyze complex data effectively.

4. Knowledge Dissemination and Impact: Identify key stakeholders who will benefit from the research findings. Develop clear and concise communication strategies to reach the target audience. Consider how the research findings can inform policy decisions and practical interventions.
5. Ethical Considerations: Assess the potential ethical implications of the research and adhere to ethical guidelines and regulations to ensure responsible research practices.

By applying systemic thinking to research planning, researchers can:

- Identify More Relevant Research Questions: Consider the broader context and interconnectedness of various factors to uncover critical and impactful research questions.
- Design More Rigorous and Robust Research Studies: Develop research designs that account for the complex dynamics of systems and potential feedback loops.
- Generate More Impactful Research Findings: Produce research that addresses the root causes of problems and offers sustainable solutions.
- Contribute to Sustainable and Equitable Solutions: Design research that considers the long-term implications of findings and promotes equitable outcomes.

Job Searching: Systemic thinking, a holistic approach to problem-solving, can be a powerful tool in your job search. By considering the broader context and the interconnectedness of various factors, you can make more informed decisions and increase your chances of securing your ideal role.

Systemic thinking can significantly enhance your job search by providing a holistic perspective. Here is how you can apply this approach:

1. Define Your Ideal Job: Consider not only the job title and salary but also factors like company culture, work-life balance, and career growth opportunities. Align your job search with your long-term career goals for a more fulfilling outcome.
2. Understand the Job Market: Research industry trends and emerging job opportunities. Identify the skills and experiences that employers are seeking. Assess the competition and differentiate yourself by emphasizing your unique strengths.
3. Tailor Your Job Search Strategy: Consider how different strategies, such as networking, online applications, and referrals, can complement each other. Use feedback from interviews and rejections to refine your approach continuously.
4. Networking Effectively: Build strong relationships with people in your industry, including former colleagues, classmates, and professionals. Share information and insights, and focus on mutual benefit to cultivate meaningful connections.
5. Crafting Your Job Search Materials: Customize each application's resume and cover letter. Use storytelling techniques to highlight your relevant skills and experiences. Develop key messages you want to convey to potential employers.
6. Preparing for Interviews: Consider the interviewer's perspective and the company's needs. Practice answering common questions and seek feedback from friends, family, or a career coach to refine your responses.

By adopting a systemic approach to your job search, you can identify new opportunities, overcome challenges, and increase your chances of landing your ideal role.

3.1.2.4 Systemic Thinking for Effective Decision-Making

Systemic thinking provides a powerful framework for making well-informed and effective decisions. By understanding the interconnectedness of various elements within a system, we can anticipate potential consequences and make choices that foster long-term sustainability.

Here is why systemic thinking is essential:

1. *Holistic Perspective*: Systemic thinking encourages us to consider the broader context of a problem or issue rather than focusing solely on immediate concerns. By taking a holistic perspective, we can avoid short-sighted solutions and make choices that promote sustainable outcomes.
2. *Identifying Root Causes*: Systemic thinking helps us delve deeper into problems to identify underlying causes. By addressing root causes, we can implement more effective and sustainable solutions.
3. *Anticipating Unintended Consequences*: Systemic thinking allows us to foresee potential unintended consequences of our decisions. By considering the broader context, we can take steps to minimize negative impacts and maximize positive outcomes.
4. *Promoting Innovation and Creativity*: Systemic thinking encourages us to challenge assumptions, think critically, and explore innovative solutions. By understanding the interconnectedness of different elements, we can foster collaboration and generate creative ideas.
5. *Building Resilience*: Systemic thinking helps us identify vulnerabilities and develop strategies to build resilience. Understanding a system's dynamics allows us to respond effectively to unexpected challenges.

By adopting a systemic approach to decision-making, we can make more informed, sustainable, and effective choices that benefit individuals, organizations, and society.

3.2 System Models Highlighting Interactions and Feedback

System models emphasize the relationships between components and serve as valuable tools for understanding how these elements interact to produce system-level outcomes. These models are particularly useful for analyzing complex systems in which interdependence drives behavior.

Below are some common types of system models that emphasize relationships:

- *Network Models*: These models illustrate connections between elements by representing them as nodes and edges in diagrams. They are commonly used for social networks, communication networks, transportation networks, and more.
- *Relational Models*: These models highlight structured relationships between data elements, such as customer-order or product-category relationships. They are often visualized using entity-relationship (ER) diagrams.
- *Dependency Models*: These models showcase hierarchical or causal relationships between elements, such as software dependencies or supply chain networks. Typically, they are visualized as directed acyclic graphs (DAGs).
- *Interaction Models*: These models focus on dynamic interactions between elements over time. Examples include agent-based models and system dynamics models, which are often visualized using flowcharts and state diagrams.
- *Bayesian Networks*: These models emphasize probabilistic relationships between elements, making them particularly effective for applications like medical diagnosis and risk assessment. They are visualized as directed acyclic graphs combined with conditional probabilities.

In the following, we will explore two specific models: quantitative input–output models and qualitative causal loop diagrams.

3.2.1 Input–Output Models

Leontief (1936, 1986) introduced input–output analysis, a groundbreaking empirical approach to studying interindustry relationships within an economy. This method provided a practical framework for analyzing economic structures, complementing theoretical models of general equilibrium. In recognition of his profound contributions to economic modeling and decision-making, Leontief was awarded the Nobel Prize in Economics in 1973.

An *input–output model* is a quantitative framework that illustrates the interdependencies between different sectors of an economy. It highlights how the output of one sector (e.g., steel production) becomes an input for another (e.g., automobile manufacturing), providing a detailed picture of the flow of goods and services within the economy. Policymakers and analysts use this model to assess economic performance, predict the effects of changes in demand, and develop strategies for sustainable growth.

Table 3.2 outlines the general structure of an input–output table, which organizes data into intermediate demand, final demand, imports, and total output. Note that the following relationship holds for each sector: *Total Input equals Total Output, which represents the total production in the economy.*

Below is a description of the symbols in Table 3.2:

- X_{ij}: Output of sector i used as input by sector j
- B_j: Value-added (e.g., employee income, business surplus, capital depreciation)
- F_i: Final demand (e.g., consumption, investment, export)

Table 3.2 General structure of an input–output table

Sector	intermediate demand					Final demand	Import	Total output
	Sector 1	…	Sector i	…	Sector n			
Sector 1	X_{11}	…	X_{1j}	…	X_{1n}	F_1	M_1	X_1
⋮	⋮		⋮		⋮	⋮	⋮	⋮
Sector i	X_{i1}	…	X_{ij}	…	X_{in}	F_i	M_i	X_i
⋮	⋮		⋮		⋮	⋮ F_n	⋮ M_n	⋮ X_n
Sector n	X_{n1}	…	X_{nj}	…	X_{nn}			
Value added	B_1	…	B_j	…	B_n			
Total input	X_1	…	X_j	…	X_n			

- M_i: Imports
- X_i: Total output (production amount)

The following equations describe the relationships:

1. Total input (supply) X_j to sector j:

$$X_j = \sum_{i=1}^{n} X_{ij} + B_j$$

2. Total output (production) X_i of sector i:

$$X_i + M_i = \sum_{j=1}^{n} X_{ij} + F_i$$

3. Input coefficient a_{ij}:

$$a_{ij} = \frac{X_{ij}}{X_j}$$

Using matrix notation, the production equilibrium for all sectors can be expressed as:

$$X + M = AX + F$$

Here:

$$X = \begin{pmatrix} X_1 \\ \vdots \\ X_n \end{pmatrix}, \ M = \begin{pmatrix} M_1 \\ \vdots \\ M_n \end{pmatrix}, \ F = \begin{pmatrix} F_1 \\ \vdots \\ F_n \end{pmatrix}, \ A = \begin{pmatrix} a_{11} & \cdots & a_{1n} \\ \vdots & \ddots & \vdots \\ a_{n1} & \cdots & a_{nn} \end{pmatrix}$$

Rearranging gives:

$$X = (I - A)^{-1}(F - M)$$

where I is the identity matrix.

The *Leontief inverse* $(I - A)^{-1}$ represents the total input requirements per unit of final demand, including both direct and indirect contributions. This inverse matrix expands as follows:

$$(I - A)^{-1} = I + A + A^2 + A^3 + \cdots$$

This expansion illustrates how changes in final demand propagate throughout the economy, capturing ripple effects across multiple sectors. This expansion is valid when the spectral radius of A is less than 1, ensuring convergence. The spectral radius of A is the eigenvalue of A with the largest absolute value.

In the input–output framework, GDP is expressed as:

$$GDP = \sum_{j=1}^{n} B_j = \sum_{i=1}^{n} (F_i - M_i)$$

This formula represents total domestic value added, ensuring that only domestically produced goods and services contribute to GDP by subtracting imports from final demand.

3.2.1.1 A Simplified Example of Input–Output Analysis

To better understand the core concepts of input–output models, consider a simplified example adapted from information provided by the Ministry of Internal Affairs and Communications of Japan.[1] This example uses a basic input–output table with just two industries, as shown in Table 3.3. The table shows how each industry's output is used by other industries (Intermediate Demand) and final consumers (Final Demand).

We can analyze Industry P as follows (Industry Q can be analyzed similarly):

- Column-wise (Inputs):
  - Industry P purchases \$30 million worth of inputs from itself (Industry P).
  - It purchases \$60 million worth from Industry Q.

[1] https://www.soumu.go.jp/toukei_toukatsu/data/io/system.htm

Table 3.3 A simplified basic input–output table (Unit: $1 million)

	Intermediate demand		Final demand	Total output
	Industry P	Industry Q		
Industry P	$X_{11} = 30$	$X_{12} = 150$	$F_1 = 120$	$X_1 = 300$
Industry Q	$X_{21} = 60$	$X_{22} = 250$	$F_2 = 190$	$X_2 = 500$
Value added	$B_1 = 210$	$B_2 = 100$		
Total input	$X_1 = 300$	$X_2 = 500$		

  - It adds $210 million in value added (e.g., labor and capital) to produce a total input of $300 million, which equals a total output.

- Row-wise (Outputs):
  - Of Industry P's $300 million total output, $30 million is sold as inputs to Industry P, and $150 million is sold to Industry Q.
  - The remaining $120 million is sold as final demand (e.g., to consumers or for exports).

Input coefficients represent the proportion of inputs used relative to an industry's total output. These coefficients are calculated by dividing each input amount by the total output of the respective industry. For example:

- Industry P's input coefficient for inputs from itself: $a_{11} = X_{11}/X_1 = 30/300 = 0.1$
- Industry Q's input coefficient for inputs from itself: $a_{22} = X_{22}/X_2 = 250/500 = 0.5$

The coefficients are summarized in Table 3.4.

The value-added coefficient for each industry is calculated as the proportion of value added to total output. For example:

- Industry P: Value-Added Coefficient $= B_1/X_1 = 210/300 = 0.7$

Input–output analysis helps us understand how changes in one industry affect others due to these interdependencies. By analyzing these flows, we can model the impact of changes in final demand on the entire economy.

Production Spillovers—Understanding Ripple Effects in Industry: Production spillovers occur when an increase in demand in one industry ripples through the economy, stimulating production in other sectors. This phenomenon highlights the interconnectedness of industries within an economy.

Table 3.4 Input coefficient table

	Industry P	Industry Q
Industry P	$a_{11} = \frac{X_{11}}{X_1} = \frac{30}{300} = 0.1$	$a_{12} = \frac{X_{12}}{X_2} = \frac{150}{500} = 0.3$
Industry Q	$a_{21} = \frac{X_{21}}{X_1} = \frac{60}{300} = 0.2$	$a_{22} = \frac{X_{22}}{X_2} = \frac{250}{500} = 0.5$

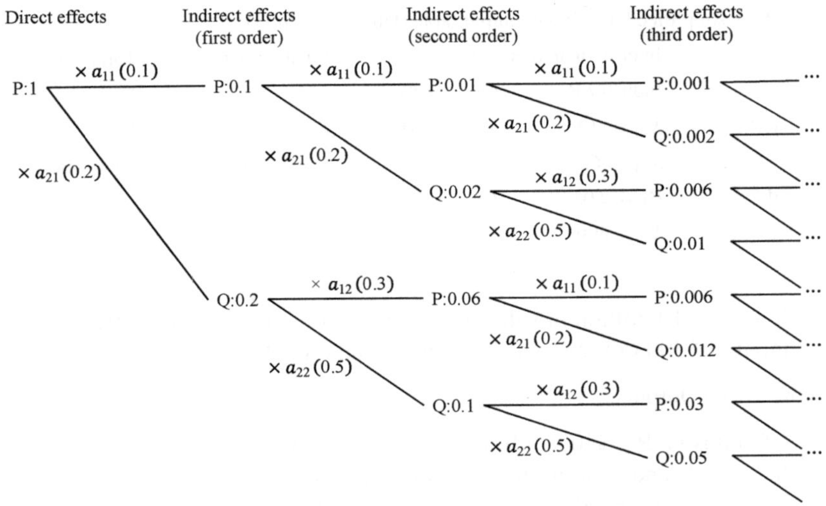

Fig. 3.1 Visualizing production spillover

As illustrated in Fig. 3.1, suppose there is a one-unit increase in demand for Industry P's products. The impact unfolds in multiple stages:

1. *Direct Effect*: Industry P increases production by 1 unit to meet the new demand.
2. *Primary Indirect Effect*: To produce this additional output, Industry P requires more inputs:
    - 0.1 units from itself.
    - 0.2 units from Industry Q.
3. *Secondary Indirect Effect*: The increased production in Industry Q creates further demand for inputs, leading to additional production increases in both industries.
4. *Recursive Ripple Effect*: This process continues in diminishing cycles, where each round of production generates new input requirements until the effects stabilize.

The total impact on Industry P can be estimated by summing the direct and indirect effects:

$$1 + 0.1 + (0.01 + 0.06) + (0.001 + 0.006 + 0.006 + 0.03) + \cdots = p_\infty$$

Similarly, the total impact on Industry Q is:

$$0.2 + (0.02 + 0.1) + (0.002 + 0.01 + 0.012 + 0.05) + \cdots = q_\infty$$

This infinite summation represents the recursive nature of spillover effects. While the series converges, it is impractical to compute directly by summing individual terms. Instead, the total impact can be efficiently determined using the inverse matrix $(I - A)^{-1}$, which captures all rounds of direct and indirect effects simultaneously.

Using the Inverse Matrix to Compute Total Impact: Given the following matrices:

$$A = \begin{pmatrix} 0.1 & 0.3 \\ 0.2 & 0.5 \end{pmatrix}, \ I - A = \begin{pmatrix} 0.9 & -0.3 \\ -0.2 & 0.5 \end{pmatrix}, \ (I - A)^{-1} = \begin{pmatrix} 1.282 & 0.769 \\ 0.513 & 2.308 \end{pmatrix}$$

For a one-unit increase in demand for Industry P:

$$F = \begin{pmatrix} 1 \\ 0 \end{pmatrix}, \ M = \begin{pmatrix} 0 \\ 0 \end{pmatrix}$$

The resulting production levels are:

$$X = (I - A)^{-1}(F - M) = \begin{pmatrix} 1.282 & 0.769 \\ 0.513 & 2.308 \end{pmatrix} \begin{pmatrix} 1 \\ 0 \end{pmatrix} = \begin{pmatrix} 1.282 \\ 0.513 \end{pmatrix}$$

Thus, for Industry P, the total production impact is $p_\infty = 1.282$, while the corresponding impact on Industry Q is $q_\infty = 0.513$.

Similarly, for a one-unit increase in demand for Industry Q:

$$F = \begin{pmatrix} 0 \\ 1 \end{pmatrix}, \ M = \begin{pmatrix} 0 \\ 0 \end{pmatrix}$$

The resulting production levels are:

$$X = (I - A)^{-1}(F - M) = \begin{pmatrix} 1.282 & 0.769 \\ 0.513 & 2.308 \end{pmatrix} \begin{pmatrix} 0 \\ 1 \end{pmatrix} = \begin{pmatrix} 0.769 \\ 2.308 \end{pmatrix}$$

This gives the total impacts of a one-unit increase in demand for Industry Q's products.

As shown in Table 3.5, these results align with the inverse matrix coefficients, which summarize the total impact of a one-unit demand increase.

This table provides a breakdown of the production increases required across industries to meet a one-unit increase in demand for a specific industry.

From this analysis, we derive three key insights:

- The inverse matrix coefficients capture both direct and indirect effects of production increases.

Table 3.5 Inverse matrix coefficients

	Industry P	Industry Q
Industry P	1.282	0.769
Industry Q	0.513	2.308

- The recursive nature of spillovers can be modeled as an infinite summation, but the inverse matrix provides an efficient way to compute the total impact.
- The column sums in Table 3.5 reveal the total economic impact of a one-unit increase in demand for each industry, demonstrating the interdependence of sectors within the economy.

3.2.1.2 Applications of Input–Output Analysis

Input–output analysis is a powerful and versatile tool that supports decision-making across various domains. Below are some key applications:

1. Economic Planning
   - Identifying Key Industries: Pinpoints sectors crucial to the economy, guiding resource allocation and investment decisions.
   - Policy Impact Assessment: Evaluates the effects of proposed policy changes, facilitating informed decision-making.
   - Economic Forecasting: Provides insights into future growth and development trends, aiding in strategic planning.
2. Business Decision-Making
   - Supply Chain Optimization: Identifies bottlenecks in supply chains and suggests strategies to improve efficiency.
   - Price Impact Assessment: Examines the effects of price changes on the economy, helping businesses adapt to market fluctuations.
   - Strategic Planning: Assesses business risks and opportunities, aiding in developing robust business strategies.
3. Environmental Decision-Making
   - Environmental Impact Assessment: Evaluates the ecological effects of economic activities to inform environmental policies and regulations.
   - Sustainable Development: Supports the design of strategies that balance economic growth with environmental protection.
4. Other Applications
   - Resource Allocation: Ensures efficient distribution of resources across different sectors.
   - Regional Development Planning: Supports strategies that promote balanced economic growth in specific regions.
   - Program Evaluation: Assesses the effectiveness of economic development programs, informing future policy improvements.

Input–output analysis is essential for understanding the intricate interconnections among various economic sectors. By leveraging this methodology, decision-makers can formulate policies and strategies that promote sustainable economic growth while addressing environmental and business challenges.

3.2.2 Causal Loop Diagrams

A *causal loop diagram* is a visual tool to represent a system's interconnected relationships and feedback loops. It helps identify the cause-and-effect relationships that influence system behavior.

The key components of a causal loop diagram are:

- *Variables*: The fundamental elements of the system, represented by circles or boxes.
- *Relationships*: The connections between variables, depicted as arrows that indicate the direction of influence (positive or negative).
- *Loops*: Circular patterns of relationships that form feedback loops, which can either reinforce or balance the system's behavior.

3.2.2.1 Basic Elements for Drawing Causal Loop Diagrams

The fundamental components of causal loop diagrams are "causal links" and "causal loops." Fig. 3.2 illustrates the two types of causal links.

In Fig. 3.2, the signs indicate the type of relationship between two variables:

- *Positive Causal Link* (+): When one variable increases (or decreases), the other also increases (or decreases), assuming all other factors remain constant. In an accumulation process, the value of the left-side variable is added to the right-side variable.
- *Negative Causal Link* (−): When one variable increases (or decreases), the other decreases (or increases), assuming all other factors remain constant. In an accumulation process, the value of the left-side variable is subtracted from the right-side variable.

As shown in Fig. 3.3, there are two main types of feedback loops:

- *Positive Feedback Loops* (Reinforcing Loops): These loops amplify changes, leading to exponential growth or decline.
- *Negative Feedback Loops* (Balancing Loops): These loops stabilize the system, promoting equilibrium and maintaining a steady state.

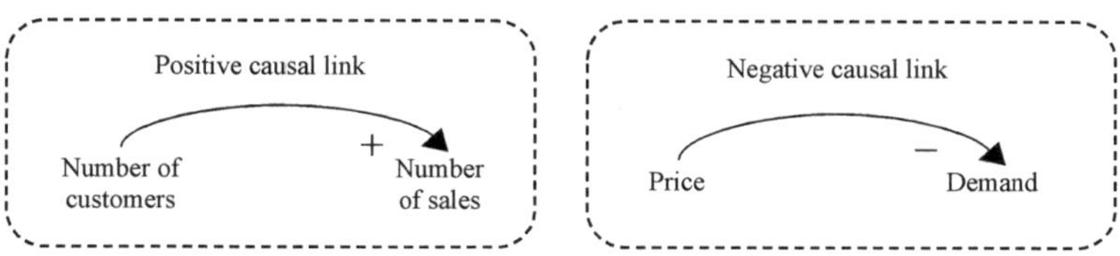

Fig. 3.2 Positive and negative causal links

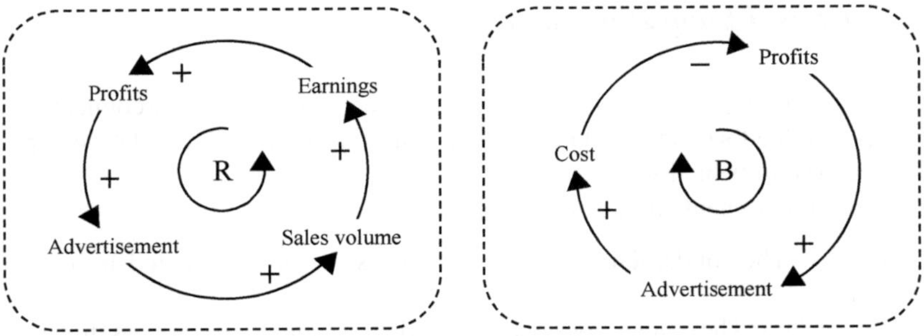

Fig. 3.3 Reinforcing (R) and balancing (B) loops

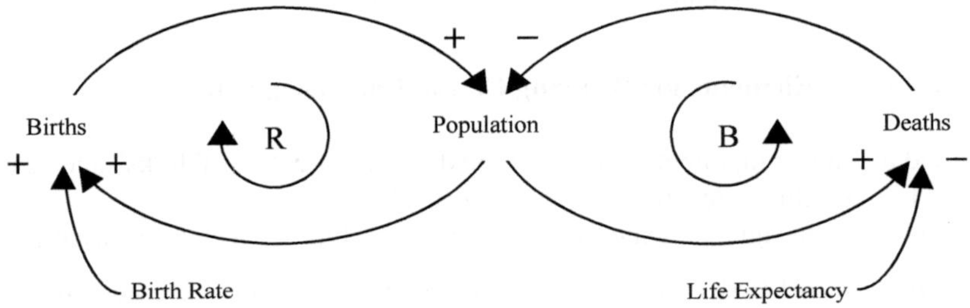

Fig. 3.4 A causal loop diagram for population growth

Figure 3.4 illustrates a simple example of population growth (Sterman, 2000), incorporating positive and negative causal links and positive and negative feedback loops.

3.2.2.2 Relationship Between Causal Loop Diagrams and System Behavior

When constructing a causal loop diagram, it is essential to understand how its elements influence system behavior. Below, we explore the connection between three key concepts in causal loop diagrams and system behavior:

- *Structure*: The underlying structure of a system plays a crucial role in shaping its behavior.
- *Time Lag*: Delays in information flow or material processes can lead to unexpected system dynamics.
- *Feedback Loops*: Both positive and negative feedback loops can amplify or attenuate changes within the system.

Sterman (2000) identifies the following six relationships between diagram elements and system behavior:

Exponential Growth: Positive feedback loops drive exponential growth when a system exhibits a constant positive growth rate. This behavior can be observed in phenomena such as compound interest and population growth. Pure exponential growth is defined by a constant doubling time, whereas a negative growth rate results in exponential decay. As shown in Fig. 3.5, the system's current state determines the net increase, which in turn further increases the system's state.

Goal-Seeking: Goal-seeking behavior occurs when a system gradually converges toward a specific target value over time. The net change in a variable is proportional to the difference between its current value and the desired goal. As shown in Fig. 3.6, this structure is typically found in balancing feedback loops. The system compares its current state to the target, and any discrepancy triggers corrective action to move the system closer to the goal.

Oscillation: Time delays in a goal-seeking system can lead to oscillatory behavior. When a delay occurs in the feedback loop, corrective actions may overshoot the target, prompting further adjustments in the opposite direction. This cyclical process can result in sustained oscillations around the goal. As shown in Fig. 3.7, delays in any of the three links within a balancing loop can cause oscillations.

S-Shaped Growth: No quantity can increase or decrease indefinitely; growth or decline eventually stabilizes. A typical system behavior is S-shaped growth, as shown

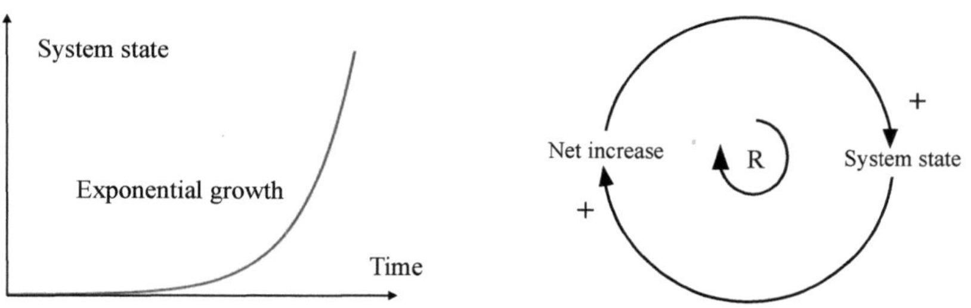

Fig. 3.5 System structure behind exponential growth

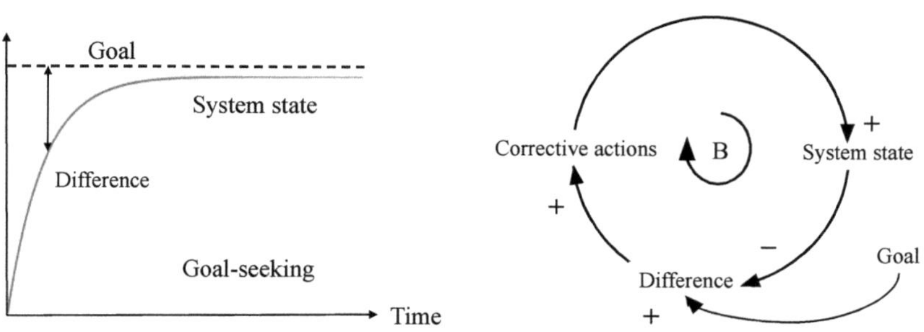

Fig. 3.6 System structure behind goal-seeking behavior

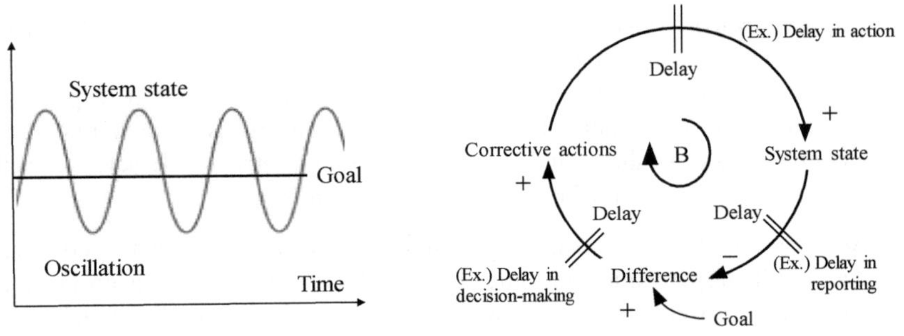

Fig. 3.7 System structure behind oscillation

in Fig. 3.8. This pattern begins with exponential growth, followed by a gradual convergence toward a steady-state value. S-shaped growth often characterizes the diffusion of innovations, such as new products or technologies, which initially spread rapidly but eventually saturate the market.

For a system to exhibit S-shaped growth, two conditions must be met:

1. Minimal Time Delays in Balancing Loops: Excessive delays can result in oscillatory behavior instead of smooth S-shaped growth.
2. Stable Carrying Capacity: The system's carrying capacity must remain relatively constant, ensuring that environmental resources are not significantly depleted.

Growth with Overshoot: If a system's limiting feedback loops experience substantial delays, it may exceed its carrying capacity and oscillate around it, as illustrated in Fig. 3.9. Timely activation of balancing feedback loops is crucial to preventing overshoot and ensuring stable growth.

Growth with Collapse: A fundamental assumption of S-shaped growth is that carrying capacity remains constant. However, human activities often deplete resources and degrade the environment, gradually reducing the system's carrying capacity. This decline introduces a second balancing loop that restricts growth, as

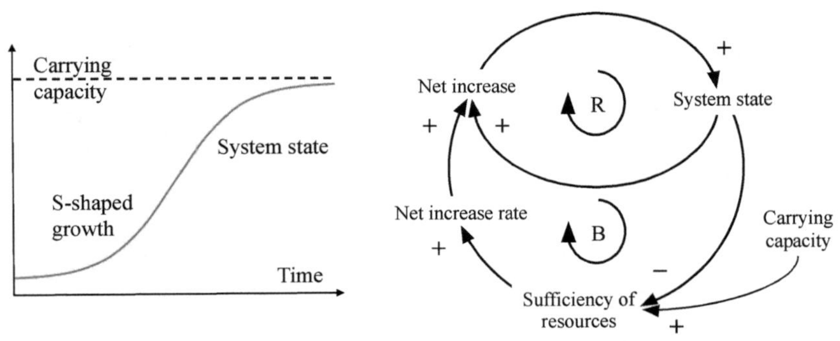

Fig. 3.8 System structure behind S-shaped growth

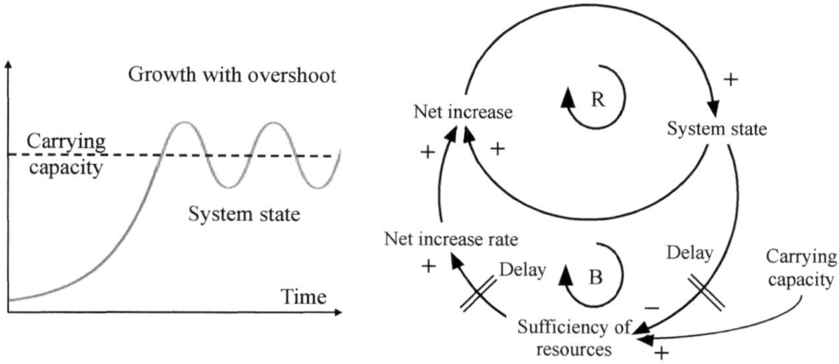

Fig. 3.9 System structure behind growth with overshoot

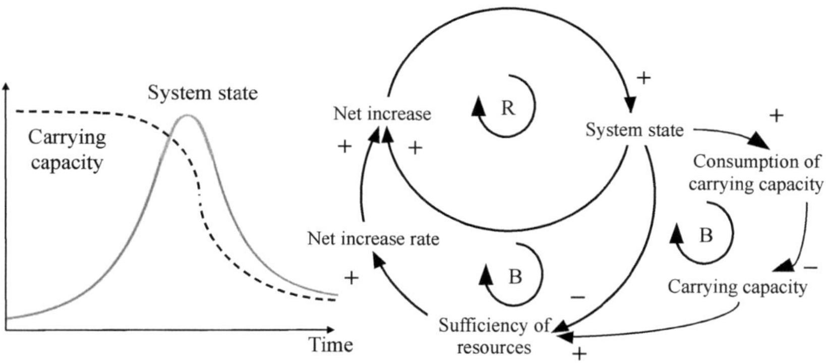

Fig. 3.10 System structure behind growth with collapse

illustrated in Fig. 3.10. In such cases, the system may initially follow an S-shaped growth pattern but eventually collapse as carrying capacity diminishes.

3.2.2.3 How Causal Loop Diagrams Inform Decision-Making

Causal loop diagrams (CLDs) are powerful tools for understanding complex systems and supporting decision-making. By visually representing the interconnections between variables, they provide a holistic perspective on how changes in one part of a system influence others. Here are some ways in which CLDs inform decision-making:

- Identifying System Dynamics: CLDs help uncover positive and negative feedback loops, providing deeper insights into system behavior. By visualizing interdependencies, they reveal behavioral patterns and highlight possible unintended consequences of interventions.

- Anticipating Consequences: CLDs enable decision-makers to explore various scenarios, predict intervention outcomes, assess risks and benefits, and identify bottlenecks and opportunities for improving system performance.
- Designing Effective Interventions: CLDs highlight key variables influencing desired outcomes, aiding in the development of effective and efficient interventions. They also help anticipate unintended consequences and devise mitigation strategies to address them proactively.
- Facilitating Collaboration: By providing a visual representation of systems, CLDs encourage discussions and foster consensus among stakeholders. They promote a systems-thinking approach, emphasizing interconnections to avoid short-term solutions that could cause long-term negative effects.

In summary, causal loop diagrams are invaluable tools for understanding complex systems, anticipating consequences, designing effective interventions, and fostering collaboration. By using CLDs effectively, decision-makers can make more informed, holistic, and effective choices.

3.2.2.4 Benefits and Challenges of Causal Loop Diagrams

Causal Loop Diagrams (CLDs) are powerful tools for understanding complex systems. They visually depict relationships between variables and feedback loops, offering several benefits, as outlined in Table 3.6. Despite their advantages, CLDs also have limitations, as summarized in Table 3.7.

To address these challenges, consider the following strategies:

- Start with Simplicity: Begin with a simplified CLD and gradually introduce complexity as needed.
- Iterative Process: Continuously refine and update the CLD as new insights emerge.

Table 3.6 Benefits of causal loop diagrams

Benefit	Description
Holistic perspective	Provides a comprehensive view of the entire system, including its components, interactions, and feedback loops. Clarifies how changes in one part affect others
Identification of root causes	Maps cause-and-effect relationships in a structured way, enabling the identification of root causes of problems and key drivers of desired outcomes
Uncovering hidden dynamics	Reveals hidden dynamics and unintended consequences that are not apparent from a linear perspective
Communication and collaboration	Serves as a shared language for discussing complex issues, enhancing collaboration among stakeholders
Systems thinking	Encourages an approach that emphasizes interconnectedness and the importance of considering the whole system

Table 3.7 Challenges of causal loop diagrams

Challenge	Description
Complexity	Leads to intricate CLDs that are difficult to understand and interpret, especially in complex systems
Subjectivity	Requires subjective judgments about relationships, which may lead to differences in interpretation
Limited quantification	Lacks quantitative information about the strength of relationships or the magnitude of changes
Time-consuming	Demands significant time investment, especially for large and complex systems
Skill requirement	Requires a strong understanding of systems-thinking principles and feedback loop modeling skills for effective use

- Group Collaboration: Involve diverse stakeholders to gain multiple perspectives and foster consensus during the development process.
- Quantitative Analysis: Combine CLDs with quantitative modeling techniques to enhance precision and better understand system behavior.
- Education and Training: Invest in training to develop the skills needed to create and interpret CLDs effectively.

3.2.2.5 System Dynamics and Causal Loop Diagrams

Causal loop diagrams (CLDs) are a foundational tool in system dynamics, a methodology for analyzing the interactions between a system's components over time (Forrester, 1961, 1969, 1971). CLDs visually represent the interconnected relationships and feedback loops that drive system behavior.

Within this approach, CLDs serve as an essential tool for:

- Identifying Feedback Loops: Distinguishing between reinforcing (positive) and balancing (negative) feedback loops.
- Understanding System Behavior: Tracing relationships between variables to anticipate system responses.
- Designing Interventions: Analyzing underlying dynamics to develop strategies for enhancing system performance.

Key concepts in system dynamics closely related to causal loop diagrams include:

- Stocks and Flows: Representing the accumulation and movement of quantities over time, often visualized using stock and flow diagrams.
- Feedback Loops: Positive and negative feedback loops drive system behavior, shaping its dynamics and stability.
- System Archetypes: Recurring patterns of system behavior that can be identified and analyzed using CLDs.

Feedback loops—both positive and negative—play a fundamental role in shaping system behavior. CLDs make these loops explicit, enabling a deeper understanding of their impact on system stability and change.

As a core component of system dynamics, CLDs provide a structured way to visualize and analyze system behavior, helping decision-makers anticipate unintended consequences and refine interventions.

Implementing System Dynamics: A system dynamics model represents a system using stocks, flows, and feedback loops to simulate its behavior over time. To implement such a model, follow these key steps:

1. *Identify the System*: Define the system's boundaries, including relevant variables and external factors influencing its behavior.
2. *Develop a Causal Loop Diagram*: Visualize relationships between variables using circles and arrows, and identify feedback loops that amplify or dampen changes.
3. *Create a Stock and Flow Diagram*: Represent accumulations (stocks) as rectangles and flows between them as arrows, capturing dynamic changes over time.
4. *Assign Equations*: Quantify relationships between variables using mathematical equations, ensuring unit consistency.
5. *Enter Data*: Set initial values for stocks and variables and define constant parameters.
6. *Simulate the System*: Use system dynamics software to calculate variable values over time and generate insights into system behavior.
7. *Validate and Refine*: Compare simulation results with real-world data and adjust the model as needed to enhance accuracy and predictive power.

System dynamics gained prominence through the groundbreaking book *Limits to Growth* (Meadows et al., 1972), commissioned by the Club of Rome. This work introduced the World3 computer simulation model, which explored the long-term implications of factors such as population growth, industrialization, resource depletion, pollution, and food production.

Key findings from Limits to Growth include:

- Business-as-Usual (BAU) Scenario: The model projected a potential global collapse within the twenty-first century if humanity continues on its current path of exponential population growth, industrialization, and resource consumption. Resource depletion, environmental pollution, and food shortages could trigger this collapse.
- Alternative Scenarios: The authors explored alternatives, including reduced population growth, limited industrialization, and increased resource efficiency. While these scenarios could delay or avert a global collapse, they require significant societal and behavioral shifts.

Core concepts in Limits to Growth are:

- *Exponential Growth*: The model highlighted the unsustainable nature of exponential population growth, industrialization, and resource consumption.

- *Feedback Loops*: It identified positive and negative feedback loops that amplify or stabilize system behavior. For instance, increased pollution can reduce agricultural productivity, creating a negative feedback loop.
- *Carrying Capacity*: The model posited that Earth has a finite carrying capacity beyond which human population and industrial activity cannot be sustained indefinitely.

Impact and controversy of Limits to Growth are:

- Significant Influence: Limits to Growth has been widely influential, shaping discussions about sustainability and environmental limits.
- Criticism: The book has faced criticism regarding the accuracy of its predictions and underlying assumptions. Critics argue that technological advancements and market forces could mitigate the challenges identified in the model.

Limits to Growth serves as a cautionary tale about the potential consequences of unchecked human activity. While specific predictions have been debated, the book's core message about the interconnectedness of human systems and the need for sustainable practices remains relevant today.

3.3 Systems Methodologies for Decision-Making

Systems science has been integral to decision analysis since its inception. Initially centered on mathematical and rational techniques, the field has evolved to encompass a broader range of systematic and systemic methods. These approaches incorporate both objective data and subjective human factors. This section explores a selection of these powerful tools.

3.3.1 Decision Analysis in Systems Science

Systems science has roots in various academic disciplines, but it did not emerge as a distinct field of study until Norbert Wiener introduced cybernetics in 1948. Derived from the Greek word for "steersman," cybernetics examines the principles of control and communication in both biological and artificial systems (Wiener, 1948).

Ludwig von Bertalanffy introduced the term "system" to capture the abstract notion of wholeness. His General Systems Theory addresses problems by replacing specific concepts with more general, abstract ones (von Bertalanffy, 1968). The goal is to establish a fundamental conceptual framework without compromising the essence of the theory or problem at hand.

Checkland (1981) defines a system as a purposeful whole characterized by emergent properties, hierarchical structure, communication and control mechanisms, and adaptability to environmental pressures. This definition differs from the colloquial use of the term "system," which often leads to misunderstandings in systems thinking.

Checkland and Scholes (1990) emphasize that terms like "education system" are frequently used loosely, mistakenly implying that the underlying mechanisms automatically constitute a system. This confusion stems from conflating a plausible description of reality with reality itself. While education experts may conceptualize these mechanisms as a system, claiming they inherently form a system is inaccurate.

Systems can manifest in both tangible and abstract forms. Physical systems, such as mechanical systems, exist in the material world. Conversely, conceptual systems, like economic systems, represent abstractions of complex realities. The degree of structure in these systems varies; some problems lend themselves to well-defined structures, while others are inherently more complex and difficult to formalize.

This distinction gives rise to two primary approaches to systems thinking:

- *Hard Systems Approach*: Relies heavily on quantitative methods, statistical analysis, and mathematical modeling. It is well suited for problems with clear objectives, measurable outcomes, and readily available data.
- *Soft System Approach*: Emphasizes qualitative methods, human judgment, and stakeholder involvement. It is particularly effective for complex problems with multiple, often conflicting objectives and limited quantitative data.

Hard Systems Approach: Checkland identifies operations research and systems engineering as key examples of the hard systems approach:

- *Operations Research* (OR): Employs mathematical methods to optimize complex decision-making processes. Techniques such as linear programming, queuing theory, and game theory form the foundation of quantitative tools used in decision analysis.
- *Systems Engineering*: Emphasizes a holistic perspective on complex systems. It provides a framework for understanding the intricate interplay between system components and their influence on decision-making.

Soft Systems Approach: The *Soft Systems Methodology* (Checkland, 1981) is a prominent example of this approach. Unlike the hard systems approach, which follows a goal-oriented paradigm, the soft systems approach acknowledges that systems are perceived and interpreted differently by individuals involved in a problem. Consequently, the problem-solving process in soft systems methodology engages with human values, prioritizing a learning cycle and evolving perceptions among stakeholders.

The following explores key methods from the hard systems approach—specifically, decision trees and Bayesian decision theory—as a foundation for understanding systematic, goal-oriented decision analysis. The discussion will then shift to the soft systems methodology, emphasizing its systemic perspective on problem-solving in complex contexts.

3.3.1.1 Decision Trees

A decision tree is a visual tool for modeling and analyzing decision-making processes. It represents a series of decisions and their potential outcomes through branches and nodes. Each decision point, or node, branches out into multiple options, each representing a possible outcome. By following these branches, one can explore various decision paths and their associated consequences.

The steps to create a decision tree are as follows:

1. Define the Decision: Clearly articulate the decision to be made and the desired outcome.
2. Identify Options: List all viable options along with their potential consequences.
3. Establish Decision Criteria: Determine the relevant factors and criteria that influence the choice.
4. Construct the Tree: Create a branching diagram to visually represent the decision-making process. Include decision nodes (choices), chance nodes (uncertain outcomes), and terminal nodes (endpoints).

Example—Planning a Japanese Lunch Date Using a Decision Tree: Imagine you are planning a Japanese lunch date with your girlfriend while keeping your budget in mind. Here are a few delicious options to consider:

- Shabu-shabu: A traditional hotpot dish featuring thinly sliced meat and vegetables cooked in boiling broth. The name "shabu-shabu" is an onomatopoeic term that mimics the sound of ingredients being cooked.
- Sushi: A classic dish consisting of vinegared rice topped with fresh ingredients such as seafood, vegetables, or fruits.
- Tempura Bowl: A satisfying meal of crispy, deep-fried seafood and vegetables served over a bed of rice.
- Pork Cutlet Bowl: A hearty dish featuring a breaded and deep-fried pork cutlet served over rice.

A decision tree approach can help you select the most suitable restaurant. First, categorize restaurants based on their price point: expensive or affordable. Expensive restaurants can be further classified based on whether they are ideal for a leisurely dining experience or a quick meal. For affordable restaurants, evaluate their healthiness. Figure 3.11 illustrates this decision tree. It is worth noting that this categorization may not align with everyone's perspective, as individuals have diverse interpretations of Japanese cuisine.

Decision Trees—A Versatile Machine Learning Tool: Decision trees are powerful supervised machine learning techniques used for classification and regression tasks. Their ability to handle complex, nonlinear relationships makes them versatile tools for a wide range of predictive modeling applications.

1. *Classification Trees*

   - Objective: To categorize data into discrete classes.

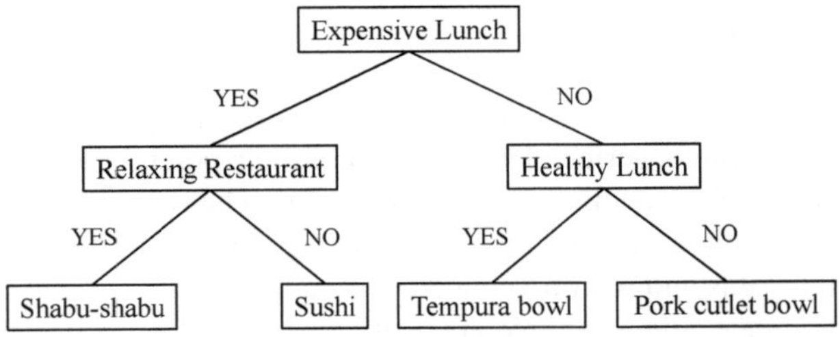

Fig. 3.11 A decision tree for selecting a restaurant

- Process: The algorithm recursively partitions the data based on feature values, creating a tree-like structure of decision rules.
- Applications: Product recommendation systems, medical diagnosis and treatment planning, and spam email filtering.

2. *Regression Trees*

- Objective: To predict continuous numerical values.
- Process: The algorithm divides the data into regions where the target variable exhibits similar values. The predicted value is the average of the target variable within each region.
- Applications: Sales forecasting, real estate price prediction, and customer churn prediction.

Unlike traditional regression models, which rely on linear relationships between variables, decision trees can effectively capture and model nonlinear patterns. Their flexibility enhances their performance on diverse and complex datasets, making them invaluable tools for both classification and regression tasks.

Algorithms for Decision Tree Creation: Several algorithms have been developed to automate decision tree creation, enabling the analysis of complex datasets. Each algorithm employs distinct methodologies and splitting criteria to optimize tree construction:

1. **CART** (Classification and Regression Trees) (Breiman et al., 1984)

- Employs a binary splitting approach, dividing data into two subsets at each node.
- Results in simple, interpretable models.
- Uses the Gini impurity as the splitting criterion.
- Requires significant computational resources for large datasets.

2. **CHAID** (Chi-squared Automatic Interaction Detection) (Kass, 1980)

- Supports multi-way splits, allowing multiple variables to be considered simultaneously.
- Particularly effective for categorical variables and uncovering complex interaction effects.

- Uses the chi-square statistic to evaluate splits.

3. **C4.5** (Quinlan, 1986)
   - Capable of performing multi-way splits.
   - Handles both categorical and numerical features.
   - Ideal for classification tasks with categorical target variables.
   - Uses the information gain ratio (based on entropy) to assess splits.

The key splitting metrics are summarized as follows:

- Gini Impurity: Measures dataset impurity; lower values indicate better splits.
- Chi-square Statistic: Evaluates the independence of variables to identify significant interactions.
- Information Gain Ratio: Quantifies entropy reduction, favoring splits with the most informative partitions.

These metrics and algorithms enable decision trees to efficiently identify patterns and relationships within datasets, tailoring the model structure to the complexity of the data.

The Gini Impurity in CART Models: The *Gini impurity*[2] is a statistical measure used in CART (Classification and Regression Trees) to evaluate the purity of a node. A lower Gini impurity indicates a purer node, where most instances belong to the same class.

The Gini impurity for a node t with c classes is calculated as:

$$G(t) = \sum_{i=1}^{c} P(i|t)(1 - P(i|t)) = 1 - \sum_{i=1}^{c} P(i|t)^2$$

where

- c: Number of classes of the target variable
- t: Current node
- $P(i|t)$: Proportion of class i in node t

In a decision tree, a Gini impurity of 0 indicates a perfectly pure node where all instances belong to the same class. The CART algorithm aims to reduce impurity by splitting nodes.

When a node t is split into two child nodes t_1 and t_2, the information gain ΔG measures the improvement in purity:

$$\Delta G(t, t_1, t_2) = G(t) - \sum_{i=1}^{2} \frac{N(t_i)}{N(t)} G(t_i)$$

[2] The term "Gini impurity" is derived from the Gini coefficient, an economic measure of inequality developed by Corrado Gini. Although the Gini impurity is conceptually different, it shares a similar mathematical foundation in measuring distributional properties.

where

- $N(t)$: Number of samples in the present node t
- $N(t_i)$: Number of samples in each child node t_i

The CART algorithm selects the split that maximizes ΔG.

Gini impurity is not used for continuous target variables. Instead, variance reduction is applied as the impurity measure. In this case, the best split is chosen to maximize the reduction in variance between the parent node and its child nodes.

Constructing decision trees involves significant calculations, typically automated using specialized software. However, to deepen your understanding of these principles, an exercise at the end of this chapter will guide you through manual calculations.

3.3.1.2 Bayesian Decision Theory

Bayesian decision theory provides a structured approach to decision-making under uncertainty, grounded in Bayes' theorem. This framework updates beliefs as new evidence emerges. The key concepts include:

- *Prior Probability*: The initial estimate of an event or hypothesis before observing data.
- *Likelihood*: The probability of observing the data given a specific hypothesis.
- *Posterior Probability*: The revised probability of an event or hypothesis after incorporating prior knowledge and observed data, calculated using Bayes' theorem.

Bayesian statistics offer several advantages:

- Incorporation of Prior Knowledge: Useful when data is limited, allowing the inclusion of existing knowledge.
- Probabilistic Results: Produces probability distributions rather than single estimates, offering a more nuanced representation of uncertainty.
- Flexibility: Adapts to a wide range of statistical problems, from simple to complex cases.
- Continuous Learning: Models improve over time by integrating new data.

Bayesian methods are widely applied across various fields, including:

- Search Engines: Refining search results based on user preferences and past interactions.
- E-commerce: Personalizing recommendations using historical behavior.
- Customer Support: Prioritizing and addressing complaints effectively.
- Artificial Intelligence: Enhancing machine learning by incorporating prior knowledge and handling uncertainty.
- Spam Detection: Improving email filtering accuracy.
- Software Development: Enhancing reliability through probabilistic modeling.

- Autonomous Vehicles: Aiding decision-making in uncertain driving conditions.
- Voice Recognition: Boosting speech recognition accuracy and natural language processing.

The Bayesian decision-making process consists of the following steps:

1. Problem Formulation: Define the decision problem, available actions, potential outcomes, and uncertainties.
2. Prior Probability Assignment: Estimate probabilities based on prior knowledge or expert opinions.
3. Data Collection: Gather relevant data for decision-making.
4. Posterior Probability Calculation: Update prior probabilities using Bayes' theorem.
5. Decision-Making: Select the action that maximizes expected utility or minimizes expected loss.

The following introduces the mathematical formulation of Bayes' theorem, followed by an application example to illustrate Bayesian decision theory in practice.

Bayes' Theorem: Given:

- H: A hypothesis or cause
- D: The observed data or result

Key terms:

- Prior Probability $P(H)$: The probability that hypothesis H is true before considering the data
- Probability of Data Occurrence $P(D)$: The probability of observing data D, regardless of H
- Likelihood $P(D|H)$: The probability of observing data D given H
- Posterior Probability $P(H|D)$: The probability that H is true given the observed data D.

Bayes' theorem calculates the posterior probability:

$$P(H|D) = \frac{P(D|H)P(H)}{P(D)}$$

This equation is derived from the multiplication rule of probability:

$$P(H \cap D) = P(H)P(D|H) = P(D)P(H|D)$$

Example—Flu or Cold? A doctor diagnosing a patient with fever and cough considers two possible conditions: flu or common cold. The patient's condition, H, can be either the flu or a cold.

Based on historical data, prior probabilities are:

- $P(H = \text{flu}) = 0.2$ (20% chance of flu)

- $P(H = \text{cold}) = 0.8$ (80% chance of cold)

The doctor administers a *rapid flu test*, which returns a positive result. However, the test has certain limitations:

- Sensitivity (True Positive Rate): 90% probability of a positive result, given the flu.
- Specificity (True Negative Rate): 80% probability of a negative result, given a cold.

Mathematically:

- $P(D = \text{positive}|H = \text{flu}) = 0.9$
- $P(D = \text{positive}|H = \text{cold}) = 1–\text{Specificity} = 1–P(D = \text{negative}|H = \text{cold}) = 0.2$

Using Bayes' theorem, the doctor updates the belief about the likelihood of flu given a positive test result:

$$P(H = \text{flu}|D = \text{positive}) = \frac{P(D = \text{positive}|H = \text{flu})P(H = \text{flu})}{P(D = \text{positive})}$$

$$= \frac{0.9 \times 0.2}{P(D = \text{positive})}$$

$$P(H = \text{cold}|D = \text{positive}) = \frac{P(D = \text{positive}|H = \text{cold})P(H = \text{cold})}{P(D = \text{positive})}$$

$$= \frac{0.2 \times 0.8}{P(D = \text{positive})}$$

The odds ratio of having the flu compared to a cold after the positive test result is:

$$\text{Odds of flu vs. cold} = \frac{0.9 \times 0.2}{0.2 \times 0.8} = \frac{18}{16} = 1.125$$

This result corresponds to a roughly 53% probability that the patient has the flu. While the test result increases the likelihood of flu, the doctor cannot be sure based on this information alone.

Note: This example shows how Bayes' theorem updates likelihoods without requiring the exact value of $P(D = \text{positive})$, as it cancels out in the comparison. In an end-of-chapter exercise, you need to calculate the probability of data occurrence to obtain posterior probabilities.

3.3.2 Soft Systems Methodology

Peter Checkland distinguished between two primary approaches to systems thinking: *hard systems thinking* and *soft systems thinking*.

- Hard systems thinking focuses on well-defined technical problems and employs quantitative methods to optimize solutions.
- In contrast, soft systems thinking tackles complex, ill-defined problems, particularly those related to human and social factors.

Checkland argued that while hard systems thinking is effective for technical challenges, it often fails to address intricate social and organizational issues. Soft systems thinking complements this by recognizing the subjective nature of problems and emphasizing the importance of human values and perceptions.

One of the most widely adopted frameworks within soft systems thinking is *Soft Systems Methodology* (SSM), developed by Peter Checkland and his colleagues (Checkland, 1981; Checkland & Scholes, 1990). SSM aims to improve real-world systems by addressing complex and ill-defined problems. It follows key principles, such as:

- *Systemically desirable*: Benefiting all relevant stakeholders.
- *Culturally feasible*: Practical to implement and sustain within the given cultural context.

Unlike approaches that seek complete consensus, SSM focuses on:

- Encouraging discussion and critical thinking.
- Fostering a deeper understanding of the problem.
- Exploring multiple perspectives and solutions.

The iterative and cyclical nature of SSM emphasizes continuous learning and adaptation. Its seven stages (illustrated in Fig. 3.12) align with the OUEI model (Fig. 1.1) as follows:

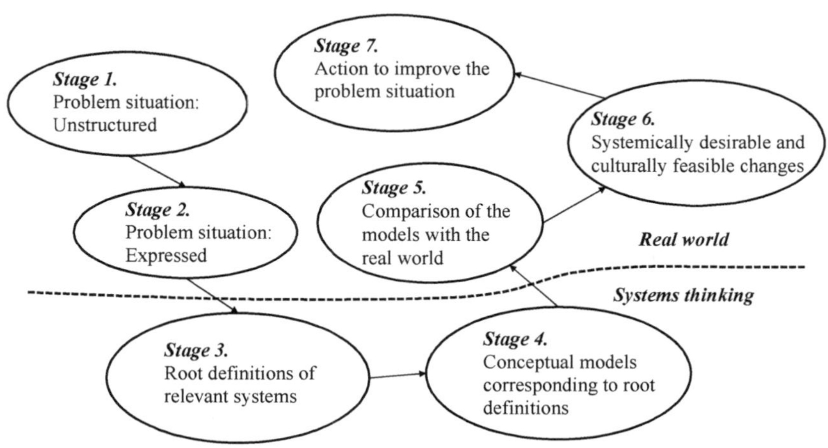

Fig. 3.12 Seven stages of the Soft Systems Methodology

- Data to Information: Stages 1 and 2 align with the data-to-information conversion phase.
- Information to Knowledge: Stages 3 and 4 correspond to the information-to-knowledge conversion phase.
- Knowledge to Ideas: Stages 5 and 6 relate to the knowledge-to-ideas phase.
- Ideas to Practice: Stage 7 aligns with the ideas-to-practice phase.

Educational Adaptations of SSM: Before exploring the seven stages of SSM through an example, consider strategies for learning SSM effectively in a classroom setting rather than through fieldwork. A classroom approach should emphasize core principles while streamlining complex steps to facilitate learning. This adaptation is not a replacement for real-world application but a structured way to help students internalize SSM's principles, enabling them to apply the methodology effectively in future professional or research settings.

Stages 3 and 4 focus on applying systems thinking to define problems and explore potential solutions. Two key tools assist in this process:

- *Root Definitions*: Define the desired transformations or changes within the system.
- *Conceptual Models*: Represent the activities and relationships required to achieve the transformations.

By comparing conceptual models to real-world situations, practitioners can uncover new insights and identify opportunities for improvement. The conceptual model is not a rigid blueprint but a tool to guide exploration of the problem space.

While SSM is a robust framework, fully implementing it, particularly Stages 4 and 5, can be time-consuming and challenging in academic settings, especially when addressing real-world problems. Therefore, a streamlined approach can be adopted to balance depth of understanding with practical constraints:

- Stage 4—Structuring the Problem Situation: Instead of developing a detailed conceptual model, students focus on identifying key system components, analyzing their inputs, processes, and outputs, and recognizing major issues. Doing this allows for a structured but manageable exploration of the system.
- Stage 5—Exploring Potential Solutions: Rather than directly comparing the conceptual model to reality, students brainstorm and discuss alternative approaches, evaluating their feasibility, implications, and alignment with the problem context.

Simplifying SSM for academic use ensures that students grasp its core concepts without becoming overwhelmed by its full complexity. However, it is essential to preserve key principles, such as:

- Understanding that conceptual models are thinking tools rather than prescriptive designs.
- Emphasizing iterative and cyclical learning, where insights evolve through discussion and refinement.

- Focusing on problem exploration rather than rushing to optimization or premature conclusions.
- Considering multiple perspectives to ensure that proposed solutions are culturally and contextually feasible.
- Encouraging case studies and group discussions as practical ways to engage with SSM in a classroom setting.

This structured yet adaptable approach allows students to develop a solid foundation in SSM, equipping them with the mindset and analytical skills necessary for future application.

3.3.2.1 Understanding SSM with a Conveyor Belt Sushi Restaurant

Imagine dining at a sushi restaurant where a continuous conveyor belt carries a variety of sushi dishes. Customers select and enjoy their favorite dishes at their own pace. Let us now examine the challenges managers face in these restaurants and explore how Soft Systems Methodology (SSM) can address them.

Innovation is essential for attracting and retaining customers in today's competitive restaurant industry. For instance, a sushi restaurant manager might develop an appealing summer menu while dealing with seasonal constraints, such as the limited availability of fresh fish caught in winter.

Stage 1—Identifying the Problem Situation: The first stage in the SSM process is recognizing and defining the problem situation. In this case, the manager observes a decline in customer numbers during summer and struggles to source high-quality fresh fish typically caught in winter. These observations reflect a vague sense of dissatisfaction—a crucial starting point for SSM analysis.

Stage 2—Expressing the Problem Situation: The second stage focuses on exploring the perceptions and perspectives of system stakeholders. This stage involves gathering insights from customers, staff, and managers to better understand their views and experiences.

For example, stakeholder feedback might include:

- Customer: "During the summer months, the restaurant should offer more than just sushi, especially for children."
- Manager: "Our menu needs to diversify, particularly in the summer when the availability of seafood that customers prefer is limited."

Understanding these diverse perspectives can reveal underlying issues and identify potential opportunities for improvement.

Rich Pictures—Visualizing the Problem Situation: Stages 1 and 2 of SSM involve identifying the problem and gathering information from various stakeholders. Rich Pictures are valuable tools for this process.

A Rich Picture is a visual representation that provides a holistic view of the system. It captures key elements, including:

- Actors (e.g., customers, staff, suppliers)
- Activities (e.g., food preparation, menu planning)
- Relationships (e.g., between managers, staff, and customers)
- Underlying Issues (e.g., seasonal challenges, customer dissatisfaction)

By creating a Rich Picture, we gain a deeper understanding of the problem context and identify key themes for further exploration. Figure 3.13 provides a simple example.

While Fig. 3.13 illustrates a basic example, Rich Pictures can be as detailed and creative as needed. The primary goal is to create a visually engaging and informative representation that sparks discussion and encourages critical thinking.

Stage 3—Developing Root Definitions: Stage 3 transitions from gathering information to constructing actionable knowledge through root definitions. These definitions represent idealized system descriptions, focusing on desired transformations or improvements.

Analyzing the diverse perspectives from Stage 2 enables the development of multiple root definitions to address identified problems and explore potential solutions. Figure 3.14 illustrates two possible root definitions for the sushi restaurant, each addressing a specific transformation, such as expanding menu options or enhancing customer satisfaction.

Using CATWOE to Structure Root Definitions: To construct robust and actionable root definitions, we employ the CATWOE technique, which systematically considers key elements of a system:

- Customers: Who benefits from the system?
- Actors: Who are the individuals or groups involved in the system?
- Transformation Process: What inputs are transformed into outputs?
- Worldview: What are the underlying assumptions and values that shape the system?

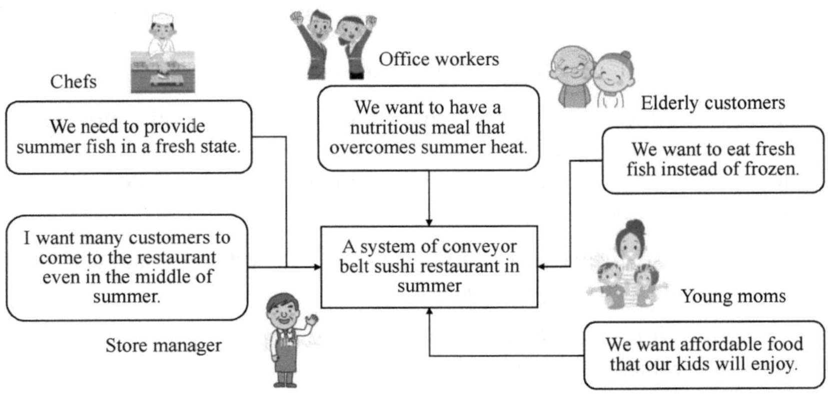

Fig. 3.13 A simple example of a Rich Picture

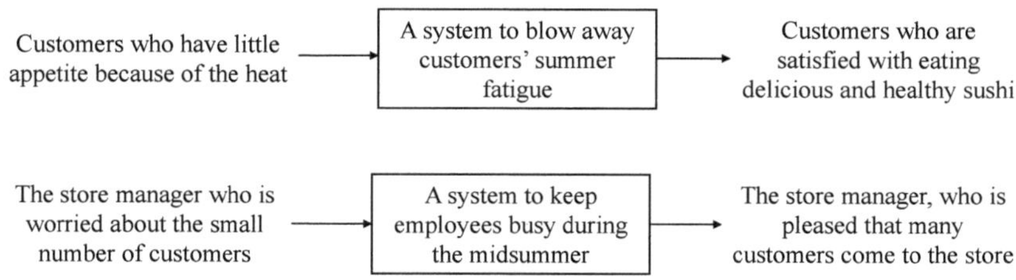

Fig. 3.14 Root definitions of an activity system

- Owner: Who has the authority to initiate or terminate the system?
- Environment: What external factors influence the system?

CATWOE provides a structured approach to ensuring root definitions are comprehensive and actionable. Using CATWOE in the sushi restaurant's context reveals the following insights:

- Customers: Restaurant patrons, especially families and older customers, may dine out less frequently during the summer.
- Actors: Restaurant staff, suppliers, and management.
- Transformation Process: Seasonal summer ingredients are transformed into appealing dishes while a steady supply of high-quality seafood, including frozen options, is maintained.
- Worldview: The restaurant should provide a comfortable and enjoyable dining experience, prioritize customer satisfaction, and operate sustainably.
- Owner: Restaurant management.
- Environment: Economic conditions, food safety regulations, and local competition.

Based on the CATWOE analysis, a possible root definition for the sushi restaurant could be:

- *A Root Definition*: A system designed to deliver a refreshing and enjoyable dining experience, especially for families and seniors during the summer months, by transforming seasonal ingredients into delicious dishes and offering a diverse menu that includes both fresh and frozen seafood options.

This root definition encapsulates the core purpose of the sushi restaurant:

- Customer Focus: Catering to families and older patrons, particularly during summer.
- Menu Diversity: Offering a range of dishes that utilize seasonal and frozen ingredients to ensure variety and quality.
- Sustainability and Adaptability: Striking a balance between customer satisfaction and operational sustainability.

Constructing this root definition clarifies the restaurant's objectives, providing a foundation for developing conceptual models and actionable strategies in subsequent stages.

Stage 4—Constructing Conceptual Models: In Stage 4, we construct a conceptual model based on a root definition, mapping the activities and relationships needed to achieve the transformations outlined in the definition. These models act as visual tools to structure discussions, evaluate potential solutions, and pinpoint areas for improvement.

In SSM, it is common to create multiple root definitions. Each root definition, reflecting unique perspectives and concerns, has a corresponding conceptual model. These models represent the system's structure and processes from their respective root definitions. By comparing and examining these different perspectives and approaches, we gain a more comprehensive understanding of the problem. This often results in multiple proposed solutions, each derived from a different root definition and conceptual model. These solutions provide a foundation for evaluating various problem-solving methods and selecting the most appropriate one.

Figure 3.15 illustrates an example of a conceptual model for a sushi restaurant, highlighting the key activities and relationships critical for addressing operational challenges. Activities such as developing a summer menu, sourcing fresh ingredients, training staff, and promoting new offerings are mapped out. By analyzing these relationships, potential bottlenecks can be identified, and strategies can be developed to optimize the process.

A conceptual model serves as a framework for solving problems. In real-world scenarios, this process must be refined through careful discussion and comparison with reality. As noted earlier, our primary goal here is to learn the Soft Systems Methodology and solve problems conceptually. Instead of strictly following the structured process shown in Fig. 3.15, we can adopt an alternative approach by focusing on key questions for consideration.

Key problems and issues in the conceptual model in Fig. 3.15 are as follows:

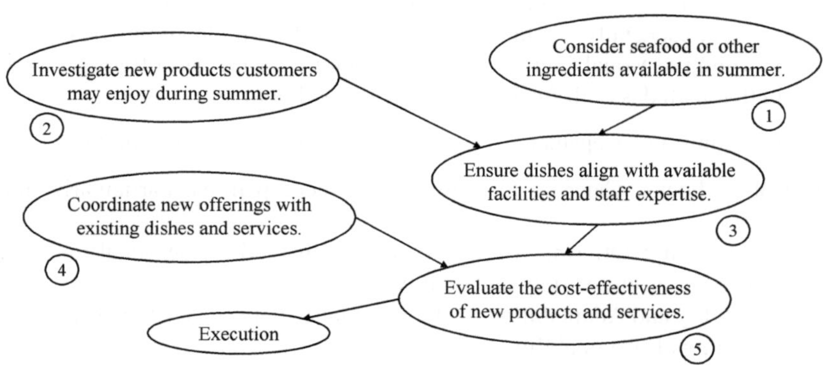

Fig. 3.15 An example of a conceptual model

1. Seasonal Ingredients: How can seafood or other summer ingredients be incorporated?
2. Product Innovation: What new products might attract customers during the summer?
3. Operational Feasibility: How can new dishes align with existing facilities and staff expertise?
4. Menu Coordination: How can new offerings complement existing dishes and services?
5. Cost Evaluation: How can the cost-effectiveness of new products and services be assessed?

Stage 5—Exploring Potential Solutions: In the original SSM process, Stage 5 compares the conceptual model created in Stage 4 with a real-world situation. However, in a simplified approach, this stage emphasizes exploring the implications of the conceptual model and transforming them into actionable proposals.

Table 3.8 presents an example of this transformation, outlining specific methods to address identified challenges and opportunities in the sushi restaurant.

The examples in Table 3.8 demonstrate how challenges can be addressed systematically using targeted solutions based on the conceptual model.

This approach transforms abstract models into actionable strategies, providing clear direction for addressing operational challenges and enhancing customer satisfaction. By focusing on seasonal ingredients, customer preferences, and resource constraints, the sushi restaurant can adopt practical strategies to improve its summer appeal while ensuring operational feasibility.

Stage 6—*Deciding on Systemically Desirable and Culturally Feasible Reforms*: In this stage, systemically desirable and culturally feasible changes are identified and implemented by assessing how proposed solutions align with the system's goals and the social and organizational context in which it operates.

Based on the analysis and discussions, the following solutions have been proposed for the sushi restaurant:

Table 3.8 Proposal for improvement of the problem situation

Challenge	Potential solution
Seasonal ingredients	Use fresh seasonal fish, such as squid, horse mackerel, conger eel, sardines, and sea urchins. Incorporate popular summer vegetables, including cucumbers, eggplants, okra, and corn, into the menu
Product innovation	Offer squid and horse mackerel sushi as highlights. Add non-sushi items such as summer vegetable salad sets, cold noodles, and chilled desserts
Operational feasibility	Expand beyond sushi by utilizing existing equipment for noodle dishes. However, hiring or training a chef for new menu items may be necessary
Menu coordination	Continue serving customer favorites like tuna, salmon, and yellowtail, even if frozen. Promote a wide variety of fresh summer seafood alongside these staples
Cost evaluation	While summer menu preparation may require considerable time and resources, promote cost-effective options like seasonal vegetables to balance expenses

- Summer Menu: Introduce a summer menu with fresh, seasonal ingredients, focusing on light and refreshing vegetable and seafood dishes.
- Cost Management: Balance the introduction of new menu items with cost-effective alternatives to maintain profitability without compromising quality.
- Customer Experience: Enhance customer satisfaction through excellent service and by creating a comfortable, inviting dining atmosphere.

Stage 7—*Implementing the Reforms*: The final stage involves implementing the agreed-upon changes and assessing their impact on the system. Although this step may be challenging to replicate in academic settings, it is crucial to consider the practical measures necessary for effective implementation.

Once the reforms are in place, their effectiveness should be monitored through feedback, performance metrics, or other evaluation tools. Continuous monitoring and adaptation are essential to ensure long-term effectiveness.

If new challenges or issues arise, the SSM process can be restarted to address them. This iterative cycle ensures continuous adaptation and improvement, fostering long-term success and sustainability.

3.3.2.2 Systemic Desirability and Cultural Feasibility

In Soft Systems Methodology (SSM), the twin concepts of systemic desirability and cultural feasibility are crucial to the success of any proposed system. These principles ensure that a system not only serves its intended purpose but also integrates smoothly into its social and organizational context.

1. Systemic Desirability: Systemic desirability refers to the extent to which a system meets its objectives and functions effectively within its environment. A system is considered systemically desirable when it:
    - Fulfills its purpose: Effectively addresses the identified problem or need.
    - Operates efficiently: Minimizes waste and optimizes resource utilization.
    - Produces intended outcomes: Achieves the desired results with measurable impact.
    - Exhibits internal coherence: Maintains a logical, well-organized structure that supports its objectives.
    - Aligns externally: Integrates well with its broader environment and complements other systems.
2. Cultural Feasibility: Cultural feasibility assesses whether a system can be successfully implemented and sustained within a specific cultural and social context. A system is considered culturally feasible when it:
    - Is understood and accepted: Stakeholders grasp its purpose and recognize its benefits.
    - Aligns with values and beliefs: Respects the norms, values, and traditions of those involved.

- Fits existing practices: Integrates seamlessly into established routines and workflows.
- Is socially acceptable: Avoids ethical or social conflicts and promotes inclusivity.
- Is sustainable: Adapts over time and remains relevant within a changing cultural context.

Balancing Desirability and Feasibility: The effectiveness of a system hinges on balancing systemic desirability and cultural feasibility.

- A system that is systemically desirable but culturally infeasible may face resistance or fail to be implemented.
- Conversely, a system that is culturally feasible but systemically undesirable might be adopted but fail to deliver meaningful outcomes.

By carefully evaluating and addressing both dimensions, stakeholders can design systems that are effective, efficient, socially acceptable, and sustainable in the long run.

3.3.2.3 Applications of SSM in Decision-Making

Soft Systems Methodology (SSM) is a flexible and robust approach for addressing complex, ill-defined problems involving multiple stakeholders. It is valuable in various decision-making contexts, ranging from strategic and policy-level decisions to operational and personal challenges.

1. *Strategic Decision-Making*: SSM helps organizations manage large-scale changes and long-term goals, such as:
    - Navigating major transitions, including mergers, acquisitions, or technological disruptions.
    - Developing comprehensive strategic plans that align with organizational objectives and stakeholder expectations.
    - Identifying systemic risks and opportunities to support sustainable growth.
2. *Policy Decision-Making*: In the public and social sectors, SSM plays a key role in shaping policies that:
    - Address societal needs while promoting equity, inclusion, and social justice.
    - Enhance the efficiency and delivery of public services.
    - Support sustainable development initiatives that consider cultural and environmental factors.
3. *Operational Decision-Making*: At the organizational level, SSM helps improve day-to-day processes by:
    - Diagnosing and resolving complex operational challenges.
    - Optimizing workflows to boost productivity and minimize inefficiencies.

- Allocating resources effectively to achieve desired outcomes.

4. *Individual Decision-Making*: On a personal level, SSM guides individuals in making thoughtful, well-informed choices, such as:
   - Planning career paths or setting personal goals.
   - Navigating relationships or ethical dilemmas by considering broader impacts.
   - Balancing priorities to align personal aspirations with societal or organizational contexts.

By promoting open dialogue, critical thinking, and a holistic systems perspective, SSM empowers decision-makers to navigate complexity and develop sustainable, adaptive solutions. Its emphasis on stakeholder engagement and iterative learning ensures that decisions are effective, contextually appropriate, and resilient over time.

3.4 Big Data and AI: Reshaping Modern Decision-Making

Big data refers to vast and complex datasets that traditional data processing tools struggle to manage effectively. These datasets are characterized by five key attributes, commonly known as the "5 Vs":

- *Volume*: The massive scale of data generated.
- *Velocity*: The speed at which data is created and processed.
- *Variety*: The diverse data formats, including structured, semi-structured, and unstructured types.
- *Veracity*: The varying levels of accuracy and reliability in the data.
- *Value*: The potential insights and benefits derived from data analysis.

The term "big data" emerged in the early twenty-first century, though no single individual or source can be definitively credited with its origin. However, Doug Laney, an analyst at Gartner, is widely recognized for popularizing the concept in 2001 with his framework of the original three Vs: Volume, Velocity, and Variety.

The scale and complexity of big data pose significant challenges for contemporary decision-making. To tackle these challenges, advanced data mining technologies organize raw information and transform it into actionable insights. At the same time, artificial intelligence (AI) is increasingly automating and enhancing decision-making processes.

This section explores the implications of these developments and examines how the integration of big data and AI is reshaping modern decision-making.

3.4.1 *Making Decisions with Big Data*

Big data plays a critical role in decision-making across various industries, typically in two primary scenarios:

A. *Immediate Decision-Making*: Once big data is structured and digitized, organizations can leverage existing knowledge to make swift, informed decisions. Examples include:

   1. Predictive Analytics: Organizations analyze new data to forecast trends using historical data and predictive models. For example, a retail company might predict future demand based on past sales, enabling timely inventory adjustments.
   2. Real-Time Monitoring and Alerts: Systems that track live data can trigger alerts based on predefined thresholds, facilitating quick responses to critical events. For instance, a network monitoring system may flag unusual traffic spikes for prompt investigation.
   3. Decision Support Systems: These systems provide tools and data to aid rapid, well-informed decisions. For example, financial analysts use such systems to assess investment opportunities by evaluating market trends and risks.
   4. Customer Segmentation: By categorizing customer data based on demographics, behavior, and preferences, organizations can personalize marketing campaigns and product offerings.
   5. Anomaly Detection: Systems designed to identify outliers or unusual patterns in data help organizations respond to potential issues. For instance, cybersecurity tools can detect suspicious activity and mitigate risks promptly.

In these cases, a strong data infrastructure combined with advanced analytical tools is essential for deriving actionable insights and making decisions quickly.

B. *Knowledge-Driven Decision-Making*: This approach involves extracting deeper insights from big data to generate new knowledge and guide decisions. Examples include:

   1. Healthcare: Analyzing patient records, genetic data, and real-time health information to develop personalized treatment plans and using real-time disease symptoms and outbreak data to predict and prevent public health crises.
   2. Finance: Detecting fraudulent activities by analyzing financial transaction patterns and evaluating creditworthiness and investment risks using historical data and market trends.
   3. Marketing: Segmenting customers based on behavior and preferences to create targeted campaigns. Predicting future customer needs to optimize product development and marketing strategies.
   4. Transportation: Using real-time traffic data to optimize traffic flow and reduce congestion and training autonomous vehicles with datasets covering road conditions, traffic patterns, and object detection.

5. Government: Analyzing population data, land use, and infrastructure needs to improve urban development. Coordinating disaster relief efforts with real-time data on weather conditions, infrastructure damage, and resource allocation.
6. Retail: Forecasting product demand based on sales data, customer behavior, and market trends. Recommending products based on customer purchase history and preferences.
7. Social Media: Analyzing posts and interactions to gauge public sentiment toward brands, products, or events and providing personalized content recommendations based on user behavior.

The success of big data initiatives depends on the effective collection, organization, and analysis of data for both immediate and knowledge-driven decision-making. By applying domain expertise and leveraging advanced tools, organizations can make informed decisions, improve efficiency, and drive innovation.

3.4.1.1 Key Data Mining Techniques

Data mining has evolved considerably, with new techniques emerging and existing ones continuously being refined. Table 3.9 summarizes some of the most prominent techniques currently used in this field.

The choice of data mining techniques depends on the specific problem, the nature of the dataset, and the desired outcomes. In practice, many real-world applications combine multiple methods to achieve optimal results.

While this section provides an overview of key methods, a solid understanding of their core principles is crucial for effective implementation. Numerous open-source tools support big data analysis, and selecting the right one depends on the specific environment and objectives.

3.4.1.2 Leveraging Big Data in Conveyor Belt Sushi Restaurants

Conveyor belt sushi restaurants must strike a delicate balance between meeting customer demand and minimizing food waste. Sushi freshness is paramount, but discarding unsold items can significantly impact profitability. To address this challenge, innovative restaurants employ big data analytics to optimize operations and reduce waste.

By embedding IC chips in sushi plates, they collect valuable data, such as:

- Popularity of Sushi Toppings: This data helps with inventory management and ingredient procurement.
- Time Spent on the Conveyor Belt: This information ensures sushi freshness, as items exceeding a certain time limit are automatically removed.
- Customer Preferences and Behavior: Analyzing customer choices and patterns helps tailor offerings and improve the customer experience.

Table 3.9 Prominent techniques used in data mining

Category	Technique	Description	References
Classification	Decision trees	A hierarchical model that splits data based on feature values to classify outcomes	Quinlan (1986), Breiman et al. (1984)
	Random forests	An ensemble method that combines multiple decision trees to improve accuracy	Breiman (2001)
	Support vector machines (SVMs)	A supervised learning model that finds a hyperplane to classify data points into categories	Vapnik (1995), Burges (1998)
	Neural networks	Computational models inspired by the human brain, suitable for complex pattern recognition tasks	Rumelhart et al. (1986), LeCun et al. (2015)
Clustering	K-means clustering	A method for partitioning data into *k* clusters by minimizing the sum of distances between data points and their assigned cluster centroids	MacQueen (1967), Bishop (2006), Hastie et al. (2009)
	Hierarchical clustering	A technique for building a hierarchy of clusters by iteratively merging similar data points	Jain et al. (1999)
	Density-based clustering	A clustering method that identifies clusters based on areas of high data density	Ester et al. (1996)
Association rule mining	A priori algorithm	An algorithm for finding frequent item sets and association rules in datasets	Agrawal and Srikant (1994)
	FP-growth algorithm	A faster approach to association rule mining, especially effective for large datasets	Han et al. (2000)
Regression	Linear regression	A modeling technique that describes relationships between dependent and independent variables as a linear equation	Draper and Smith (1981), Montgomery et al. (2021)
	Logistic regression	A predictive method for binary outcomes, such as customer churn	Hosmer and Lemeshow (2000), Agresti (2012)
	Decision trees and random forests	Tree-based models that can also be applied to regression tasks	Quinlan (1986), Breiman (2001)
Dimensionality reduction	Principal component analysis (PCA)	A technique for reducing data dimensionality while preserving key information	Pearson (1901), Hotelling (1933), Jolliffe (2002)
	t-SNE	A nonlinear method that preserves local structure in visualization tasks	Maaten and Hinton (2008)

(continued)

Table 3.9 (continued)

Category	Technique	Description	References
	Autoencoders	Neural networks trained to reconstruct input data, often used for feature extraction	Hinton and Salakhutdinov (2006), Bishop (2006)
Time series analysis	ARIMA models	A statistical technique for analyzing and forecasting time series data by combining autoregression, differencing, and moving averages	Box et al. (2015)
	LSTM networks	A type of Long Short-Term Memory network designed for handling sequential data like time series	Hochreiter and Schmidhuber (1997)
Other techniques	Ensemble methods	A strategy for combining multiple models (e.g., bagging, boosting) to improve accuracy	Breiman (1996), Freund and Schapire (1997)
	Anomaly detection	A technique for identifying data points that deviate from expected patterns	Hastie et al. (2009), Bishop (2006)
	Text mining	A method for extracting valuable information from unstructured text data	Feldman and Sanger (2007)
	Social network analysis	An approach for analyzing relationships and interactions within networks	Wasserman and Faust (1994)

By leveraging this data, restaurants can:

- Reduce Food Waste: Accurately predicting demand and optimizing inventory levels minimizes the number of unsold items.
- Improve Freshness: Tracking the time sushi spends on the conveyor belt ensures that only the freshest items are served.
- Increase Profit Margins: Reducing food waste and making data-driven decisions about ingredient procurement enhances profitability.
- Enhance Customer Satisfaction: Ensuring sushi freshness and providing diverse options delights customers and fosters loyalty.

This strategic use of big data exemplifies how technology can transform operations, reduce costs, and elevate customer satisfaction in the food service industry. By embracing innovation, conveyor belt sushi restaurants set a benchmark for blending tradition with modern analytics.

3.4.2 Autonomous Decision-Making AI

Autonomous decision-making AI refers to artificial intelligence systems designed to make independent choices or take actions without direct human intervention. These systems process information, analyze data, and reach conclusions using predefined rules, algorithms, and machine learning models.

Key characteristics of autonomous decision-making AI include:

- Self-Learning: The ability to improve decision-making by analyzing data and refining models over time.
- Real-Time Processing: Rapid analysis and decision-making in real-time scenarios.
- Adaptability: The flexibility to adjust to changing conditions and dynamically refine decision-making processes.
- Complex Problem-Solving: The capacity to address multifaceted problems involving numerous variables.

Applications of autonomous decision-making AI include:

- *Self-Driving Cars*: AI systems use sensor data to make real-time decisions about steering, braking, and acceleration.
- *Medical Diagnosis*: AI aids in diagnosing diseases by analyzing medical images, test results, and patient data.
- *Financial Trading*: AI-powered trading systems analyze market trends to autonomously make buy-and-sell decisions.
- *Customer Service Chatbots*: Intelligent chatbots interact with customers and resolve queries without human assistance.

While autonomous decision-making AI offers transformative benefits, it raises ethical and safety concerns. Key issues include:

- Accountability: Determining who is responsible for decisions made by AI systems.
- Bias: Ensuring that AI systems operate fairly and do not reinforce societal biases.
- Unintended Consequences: Mitigating the risks of unforeseen actions or decisions by autonomous systems.

Autonomous decision-making AI relies on a combination of advanced technologies:

- *Machine Learning* (ML): ML models, including deep learning and reinforcement learning, are trained on large datasets to identify patterns and make predictions. These techniques are critical for applications such as image recognition and decision-making.
- *Computer Vision*: Enables AI to interpret and understand visual data from images and videos, which is essential for facial recognition, object detection, and navigation in self-driving cars.
- *Natural Language Processing* (NLP): Allows AI to process and understand human language, supporting applications like chatbots and voice-controlled assistants.

- *Robotics*: Provides the physical components, sensors, actuators, and control algorithms needed for AI to interact with the real world.
- *Big Data Analytics*: Facilitates the analysis of massive datasets, which is crucial for training AI models and making informed decisions.
- *Edge Computing*: Processes data closer to its source, reducing latency and enhancing real-time decision-making capabilities.

Future Outlook: As AI technologies advance, autonomous systems will become increasingly sophisticated and capable of handling more complex decision-making tasks across diverse industries. These innovations have the potential to transform how we work, live, and interact with technology.

AlphaGo's Smart Decision-Making: *AlphaGo*, a groundbreaking computer Go program developed by Google DeepMind, astonished the world by defeating top human Go players between 2015 and 2017. This achievement marked a pivotal milestone in artificial intelligence and sparked a global surge in AI research and innovation.

Go is a highly complex board game that demands strategic thinking and creativity. With an astronomical number of possible board positions (approximately 2×10^{170}), it has long posed a significant challenge for traditional AI approaches reliant on brute-force search.

AlphaGo's success stemmed from its innovative use of neural networks and self-play learning:

- Neural Networks: Instead of relying on hardcoded heuristics, AlphaGo learned evaluation strategies through deep neural networks. These networks allowed it to assess board positions and predict optimal moves by identifying patterns.
- Self-Play Training: AlphaGo refined its strategies by playing millions of games against itself, improving its decision-making capabilities through iterative learning.
- Monte Carlo Tree Search: This technique enhanced AlphaGo's decision-making by efficiently exploring potential move sequences and prioritizing the ones most likely to succeed.

The significance of AlphaGo lies in its demonstration that AI can achieve superhuman performance without human-like intuition. Instead, it relied on data-driven learning and advanced algorithms to surpass even the most skilled human players. This accomplishment redefined the boundaries of AI capabilities and showcased its potential to revolutionize various domains, challenging traditional notions of human intelligence and creativity.

AlphaGo appears again in Chapter 5, where we compare machine learning with human learning.

3.5 End-of-Chapter Problems

3.5.1 Problem 3.1: Education System

Checkland and Scholes emphasize a critical distinction: while we often refer to "education systems," these do not inherently function as systems. Instead, they become systems only when we consciously conceptualize and analyze them as such. An education system represents a complex interplay of diverse elements, including institutions, policies, individuals, and societal influences.

A graduate school serves as an illustrative case study. As a subsystem within the broader education system, it may appear well defined but is influenced by numerous internal and external factors:

Internal factors include:

- Organizational Structure: Departments, programs, and administrative units.
- Faculty: Their expertise, research focus, and teaching philosophies.
- Students: Backgrounds, motivations, and learning preferences.
- Curriculum: Courses offered and academic standards maintained.
- Resources: Funding, facilities, and technological tools.

External factors include:

- Government Regulations: Policies, funding priorities, and educational standards.
- Economic Conditions: Job market demands, funding opportunities, and enrollment trends.
- Societal Trends: Demographic shifts, cultural changes, and technological advancements.
- Global Context: International collaborations, student mobility, and global rankings.

These elements interact dynamically, often in non-harmonious ways. For example, government policies may diverge from institutional goals, while faculty priorities might not always align with student needs. Addressing these complexities requires a systemic perspective.

Task: Explain how interconnectedness, holism, feedback loops, and emergent properties apply to a graduate school viewed as a system.

3.5.2 Problem 3.2: Causal Loop Diagram

Bicycle-sharing businesses have gained widespread popularity due to their convenience. However, these businesses face challenges such as intense competition, limited parking availability, and regulatory constraints.

Task: Construct a causal loop diagram focusing on the variable "Number of Bicycles Available to Rent" and analyze its impact on the system. Consider the interplay between the following variables:

- Number of Bicycles Available to Rent
- Number of Market Entrants
- Government Regulations
- Number of Users
- Quality of Urban Life
- Demand for Bicycle Parking
- Air Quality and Road Safety
- Operation Rates of Other Modes of Transport
- Revenue of Bicycle-Sharing Service Companies
- (Add additional variables as needed.)

By examining these relationships, we can predict potential outcomes and develop strategies to ensure the sustainability and growth of bicycle-sharing businesses.

3.5.3 Problem 3.3: Decision Tree

Imagine planning a whitewater rafting adventure, but the destination is far away, and you want to avoid a last-minute cancellation due to adverse weather.

Whitewater rafting is an exhilarating outdoor activity where participants navigate a river's rapids in an inflatable raft. The rapids are formed by water flowing over rocks and other obstacles, creating both thrilling challenges and potential hazards.

However, certain weather conditions can lead to the cancellation of rafting trips:

- Flooding: Rapidly rising water levels can make the river unsafe and impassable.
- Severe Weather: Strong winds, thunderstorms, and heavy rain create hazardous conditions. While light rain is generally manageable, severe weather often results in trip cancellations.
- Natural Disasters: Earthquakes or landslides can drastically alter river conditions and accessibility.

You have access to historical data on rafting operations under various weather conditions (Table 3.10). This dataset includes the water level (low, normal, high), wind strength (none/weak, strong), and rainfall (none/light, heavy).

Task: Using this information, apply a CART model to determine the weather conditions under which you should proceed with your rafting trip without relying on external sources.

Note: The small dataset increases the risk of overfitting—your decision tree may classify all 12 cases perfectly without generalizing well. However, the primary goal of this exercise is to practice using the CART model, so do not worry about overfitting for now.

Table 3.10 Weather conditions and rafting operations

Case	Water level	Wind strength	Rainfall	Operation
1	Normal	None/Weak	Heavy	No
2	High	None/Weak	None/Light	Yes
3	Low	None/Weak	Heavy	Yes
4	Normal	Strong	Heavy	No
5	Normal	Strong	None/Light	Yes
6	Low	Strong	None/Light	No
7	High	Strong	None/Light	No
8	Low	None/Weak	None/Light	Yes
9	High	None/Weak	Heavy	No
10	Low	Strong	Heavy	No
11	High	Strong	Heavy	No
12	Normal	None/Weak	None/Light	Yes

3.5.4 *Problem 3.4: Bayesian Decision Theory*

Imagine a scenario where a new infectious disease emerges, affecting 3% of the population in your city, similar to the COVID-19 pandemic.

You take an antigen test, and the result is positive. Given the following:

- 70% of infected individuals test positive (sensitivity).
- 10% of non-infected individuals test positive (false positive rate).

You can calculate the probability of actual infection in the following cases:

1. Asymptomatic Testing: You have no symptoms and took the test for reassurance.
2. Symptomatic Testing: You have symptoms and estimate an 80% chance of infection.

Task: Use Bayes' Theorem to determine the posterior probability of infection for Cases 1 and 2, given a positive test result.

3.5.5 *Problem 3.5: Soft Systems Methodology*

We will analyze the problem of designing a health management system for residents in a small city using the Soft Systems Methodology (SSM).

Stage 1—Problem Situation: The mayor of a small city is concerned about rising healthcare costs due to an aging population. He envisions a society where elderly individuals can remain active and healthy. One proposed solution is a remote health management system, particularly tailored for individuals with mild diabetes.

Stage 2—Stakeholder Perspectives: Those involved think about this issue as follows:

1. Mayor/City Officials: Aim to develop a system that reduces healthcare costs and promotes public health.
2. Hospitals/Clinics: Require a system that supports patients with mild symptoms, helping to ease the workload of healthcare professionals.
3. Sensor Development Company: Seek to implement simple sensors for monitoring blood glucose and urine sugar levels.
4. Software Development Company: Plan to leverage mobile data tools to enable seamless communication between patients and healthcare advisors.
5. Patients with Mild Diabetes: Desire a system that alleviates health-related anxieties and promotes self-management.

Task: Based on the stakeholder perspectives from Stage 2, progress to Stage 3 and beyond of the Soft Systems Methodology (SSM):

- *Stage 3—Root Definition*: Develop a concise and comprehensive definition of a remote health management system that aligns with the needs and aspirations of all stakeholders. This definition should clearly capture the essence of the system and its purpose.
- *Stage 4—Conceptual Model*: Identify the key problems or issues that must be addressed to achieve the root definition. Then, create a conceptual model—a high-level representation of the system—to visualize the relationships between different components and processes. This model should emphasize the critical factors influencing the system's effectiveness, such as technology, human factors, and organizational structures.
- *Stage 5—Potential Solutions*: Explore potential solutions for the identified problems. Consider a range of options, including technological innovations and organizational changes. Evaluate the feasibility and desirability of each solution, taking into account factors such as cost, time, and potential impact.
- *Stage 6—Systemically Desirable and Culturally Feasible Solutions*: Select promising solutions that are both technically feasible and culturally acceptable. Develop a detailed implementation plan that outlines the necessary steps, timelines, and resource requirements. Additionally, address the ethical implications of the proposed solutions.

References

Agrawal, R., & Srikant, R. (1994). Fast algorithms for mining association rules in large databases. In *VLDB '94: Proceedings of the 20th International Conference on Very Large Data Bases* (pp. 487–499).

Agresti, A. (2012). *Categorical data analysis*. Wiley.

Beer, S. (1979). *The heart of enterprise*. Wiley.

Beer, S. (1984). The viable system model: Its provenance, development, methodology and pathology. *Journal of the Operational Research Society, 35*, 7–25.
Bishop, C. M. (2006). *Pattern recognition and machine learning*. Springer.
Box, G. E. P., Jenkins, G. M., & Reinsel, G. C. (2015). *Time series analysis: forecasting and control*. Wiley.
Breiman, L. (2001). Random forests. *Machine Learning, 45*(1), 5–32.
Breiman, L., Friedman, J. H., Olshen, R. A., & Stone C. J. (1984). *Classification and regression trees*. Chapman and Hall/CRC.
Breiman, L. (1996). Bagging predictors. *Machine Learning, 24*(2), 123–140.
Brown, J., & Isaacs, D. (2005). *The world café: Shaping our futures through conversations that matter*. Berrett-Koehler Publishers.
Burges, C. J. C. (1998). A tutorial on support vector machines for pattern recognition. *Data Mining and Knowledge Discovery, 2*, 121–167.
Buzan, T., & Buzan, B. (1996). *The mind map book: How to use radiant thinking to maximize your brain's untapped potential*. Plume.
Checkland, P. B. (1981). *Systems thinking*. Wiley.
Checkland, P. B., & Scholes, J. (1990). *Soft systems methodology in action*. Wiley.
Cooperrider, D. L., & Whitney, D. (2005). *Appreciative inquiry: A positive revolution in change*. Berrett-Koehler Publishers.
Dess, G. G., Lumpkin, G. T., & Eisner, A. B. (2005). *Strategic management: Text and cases*. Irwin Professional.
Draper, N. R., & Smith, H. (1981). *Applied regression analysis* (2nd ed.). Wiley.
Ester, M., Kriegel, H.-P., Sander, J., & Xu, X. (1996). A density-based algorithm for discovering clusters in large spatial databases with noise. In KDD'96: *Proceedings of the Second International Conference on Knowledge Discovery and Data Mining* (pp. 226–231).
Feldman, R., & Sanger, J. (2007). *The text mining handbook: Advanced approaches in analyzing unstructured data*. Cambridge University Press.
Freund, Y., & Schapire, R. E. (1997). A decision-theoretic generalization of on-line learning and an application to boosting. *Journal of Computer and System Sciences, 55*(1), 119–139.
Forrester, J. W. (1961). *Industrial dynamics*. Massachusetts Institute of Technology Press.
Forrester, J. W. (1969). *Urban dynamics*. Pegasus Communications.
Forrester, J. W. (1971). *World dynamics*. Wright-Allen Press.
Han, J., Pei, J., & Yin, Y. (2000). Mining frequent patterns without candidate generation. *ACM SIGMOD Record, 29*(2), 1–12.
Hastie, T., Tibshirani, R., & Friedman, J. H. (2009). *The elements of statistical learning: Data mining, inference, and prediction* (2nd ed.). Springer.
Hinton, G. E., & Salakhutdinov, R. R. (2006). Reducing the dimensionality of data with neural networks. *Science, 313*(5786), 504–507.
Hochreiter, S., & Schmidhuber, J. (1997). Long short-term memory. *Neural Computation, 9*(8), 1735–1780.
Hosmer, D. W., & Lemeshow, S. (2000). *Applied logistic regression* (2nd ed.). Wiley.
Hotelling, H. (1933). Analysis of a complex of statistical variables into principal components. *Journal of Educational Psychology, 24*(6), 417–441.
Jain, A. K., Murty, M. N., & Flynn, P. J. (1999). Data clustering: A review. *ACM Computing Surveys, 31*(3), 264–323.
Jolliffe, I. T. (2002). *Principal component analysis*. Springer.
Kass, G. V. (1980). An exploratory technique for investigating large quantities of categorical data. *Journal of the Royal Statistical Society. Series C (Applied Statistics), 29*(2), 119–127.
LeCun, Y., Bengio, Y., & Hinton, G. (2015). Deep learning. *Nature, 521*, 436–444.
Leontief, W. (1936). Quantitative input and output relations in the economic system of the United States. *Review of Economic Statistics, 18*(3), 105–125.
Leontief, W. (1986). *Input-output economics* (2nd ed.). Oxford University Press.

van der Maaten, L. J. P., & Hinton, G. E. (2008). Visualizing data using t-SNE. *Journal of Machine Learning Research, 9*, 2579–2605.

MacQueen, J. B. (1967). Some methods for classification and analysis of multivariate observations. In *Proceedings of the FIFTH BERKELEY SYMPOSIUM on Mathematical Statistics and Probability, Volume 1: Statistics* (pp. 281–297). University of California Press.

Meadows, D. H., Meadows, D. L., Randers, J., & Behrens, W. W., III. (1972). *The limits to growth; a report for the club of Rome's project on the predicament of mankind.* Universe Books.

Montgomery, D. C., Peck, E. A., & Vining, G. G. (2021). *Introduction to linear regression analysis* (6th ed.). Wiley.

Morecroft, J. D. W. (2007). *Strategic modelling and business dynamics: A feedback systems approach.* Wiley.

Pearson, K. (1901). On lines and planes of closest fit to systems of points in space. *The London, Edinburgh, and Dublin Philosophical Magazine and Journal of Science, 2*, 559–572.

Quinlan, J. R. (1986). Induction of decision tees. *Machine Learning, 1*(1), 81–106.

Rumelhart, D. E., Hinton, G. E., & Williams, R. J. (1986). Learning representations by back-propagating errors. *Nature, 323*, 533–536.

Schwartz, P. (1996). *The art of the long view: Planning for the future in an uncertain world.* Crown Currency.

Sterman, J. D. (2000). *Business dynamics: Systems thinking and modeling for a complex world.* McGraw Hill.

Vapnik, V. N. (1995). *The nature of statistical learning theory.* Springer.

Vennix, J. (1996). *Group model building: facilitating team learning in system dynamics.* Wiley.

von Bertalanffy, L. (1968). *General systems theory.* George Braziller.

Wasserman, S., & Faust, K. (1994). *social network analysis: Methods and applications.* Cambridge University Press.

Wiener, N. (1948). *Cybernetics.* MIT Press.

Wilson, B. (1990). *Systems: Concepts, methodologies, and applications.* Wiley.

Further Reading on Systems Thinking and Methodologies

Jackson, M. C. (1991). *Systems methodology for the management (Contemporary Systems Thinking).* Plenum Press. This book outlines methodologies for system management, providing tools to apply systems thinking in organizational contexts.

Jackson, M. C. (2019). *Critical systems thinking and the management of complexity.* Wiley. Focused on critical systems thinking and complexity management, this book incorporates the latest developments in contemporary systems thinking.

Meadows, D. (2008). *Thinking in systems: A primer.* Chelsea Green. A widely recognized introduction to systems thinking, this book explains the fundamental concepts and principles of the field.

Midgley, G. (ed.) (2002). *Systems thinking* (Vol. 1–4). Sage. This comprehensive four-volume set gathers influential theoretical papers and studies, offering a profound exploration of systems thinking.

Mingers, J. (2014). *Systems thinking, critical realism and philosophy: A confluence of ideas.* Routledge. This book examines the intersection of systems thinking, critical realism, and philosophy, enhancing theoretical understanding in the field.

Richmond, B. (2000). *The "Thinking" in systems thinking: Seven essential skills.* Pegasus Communications. A practical guide introducing seven essential skills for systems thinking, this book emphasizes actionable approaches.

Salmon, P. M., et al. (2022). *Handbook of systems thinking methods.* CRC Press. A comprehensive overview of systems thinking methodologies, this book supports practical applications and problem-solving techniques.

Senge, P. M. (1990). *The fifth discipline: The art & practice of the learning organization*. Random House. This influential book explores the concept of the learning organization, emphasizing systems thinking as a tool for organizational transformation.

Sweeney, L. B., & Meadows, D. (2010). *The systems thinking playbook: Exercises to stretch and build learning and systems thinking capabilities*. Chelsea Green Publishing. This book provides exercises designed to build systems thinking skills, offering practical activities for enhancing learning and application.

Chapter 4
Decision-Making with the Creative Mind

Abstract This chapter underscores the vital role of creativity and imagination in decision-making, particularly in addressing the limitations of rational and systemic approaches. It explores how creative thinking can overcome these constraints, inspiring innovative and unconventional solutions. After introducing abduction (a form of hypothetical reasoning), the chapter presents techniques to enhance creative decision-making and explores frameworks such as knowledge management and design thinking. These frameworks highlight their potential to foster innovation, generate novel ideas, and drive effective problem-solving in dynamic contexts.

4.1 Creative Thinking

Creative thinking is a cornerstone of effective decision-making, enabling individuals to:

- *Think critically*: Challenge assumptions and consider diverse perspectives.
- *Be flexible*: Adapt to change and explore unconventional possibilities.
- *Be innovative*: Develop unique ideas and approaches.

By encouraging "outside-the-box" thinking, creative approaches uncover opportunities that traditional methods often overlook. It fosters open-mindedness, adaptability, and collaboration—qualities that are essential in a fast-changing world. Moreover, it inspires individuals and cultivates a culture of innovation and collaboration.

This section explores the role of intuition in creative thinking and introduces abduction—a form of hypothetical reasoning that complements induction and deduction.

Y. Nakamori, *Sensible Decision-Making*, https://doi.org/10.1007/978-981-96-6811-3_4

4.1.1 Intuition and Decision-Making

Language development is a critical milestone in human evolution. It facilitates the intergenerational transfer of knowledge and the emergence of a rich cultural and intellectual heritage. While biological evolution, including brain development, slowed and eventually plateaued, philosophical and societal evolution accelerated.

Wierzbicki and Nakamori (2006) describe intuition as a preverbal, holistic, and subconscious ability to process sensory signals and memories—a remnant of humanity's early evolutionary stage. Intuition encompasses imagination and the comprehensive processing of sensory input and stored memories. Unlike instincts and emotions, which are largely genetically determined, *intuition is primarily shaped by learning and experience.*

4.1.1.1 Intuition and Emotion

In Chapter 2, we discussed sensibility and emotion-based decision-making. While these concepts and intuition play significant roles in human behavior and decision-making, they operate through distinct mechanisms and serve complementary purposes.

Intuition is a rapid, unconscious form of cognition that allows individuals to make judgments or form beliefs without explicit deliberation. It generates insights, hunches, or feelings based on past experiences, learned patterns, and subconscious information processing. In decision-making, intuition:

- Acts as a cognitive shortcut, especially in time-sensitive or uncertain situations.
- Provides insights or "hunches" that complement rational analysis, though it can be subject to biases.
- Is particularly valuable in complex or ambiguous scenarios where patterns are not immediately apparent.

Emotion, on the other hand, is a complex psychophysiological response to internal or external stimuli. It encompasses physiological arousal (e.g., heart rate changes), subjective feelings (e.g., joy, sadness), expressive behaviors (e.g., smiling, frowning), and cognitive appraisals (e.g., interpreting a situation as threatening or rewarding). Emotions influence decision-making by:

- Providing motivation and guiding priorities, signaling what matters most.
- Enhancing or clouding judgment, depending on their intensity and nature (e.g., fear or anger may impair judgment, while empathy can drive ethical decisions).
- Lasting longer than intuition due to their physiological and psychological components.

Intuition involves unconscious thinking and knowing, while emotion centers on feeling and reacting. Both are fundamental to human behavior and decision-making, functioning through distinct mechanisms yet complementing each other in shaping our responses to the world.

4.1.1.2 The Importance of Intuition in Decision-Making

In an era increasingly driven by data and logic, the role of intuition in effective decision-making is often underestimated. While rational analysis provides a vital foundation, truly exceptional decision-making emerges from a seamless integration of logic and intuition.

In this context, intuition is a subconscious understanding shaped by accumulated experience and acquired knowledge. It is not merely a gut feeling but a form of deeply ingrained wisdom that transcends easily articulated facts. Michael Polanyi (1958, 1966) described this as "tacit knowledge"—an implicit understanding that profoundly influences decisions despite being difficult to express explicitly.

Beyond decision-making, intuition also fuels creativity. Techniques such as brainstorming and mind mapping allow individuals to tap into their intuitive understanding, forging novel connections between existing knowledge and memories. Like innovation, this process often involves the unexpected recombination of familiar elements, leading to original solutions.

In today's information-rich world, explicit knowledge and readily available data have become increasingly commoditized. The true differentiator lies in the unique, often unspoken expertise embedded within each individual's mind.

While logical, data-driven decision-making is indispensable, it can sometimes oversimplify reality or distort complex problems. Many real-world challenges extend beyond the scope of available data, requiring a more holistic approach that integrates both logic and intuition. These two forces are not opposing but complementary:

- Logic provides a structured framework for evaluating decisions.
- Intuition allows us to transcend the limitations of purely data-driven analysis.

For instance, doctors may use medical knowledge (logic) to assess symptoms, yet intuition might prompt them to consider less apparent diagnoses or order specific tests.

In practice, the notion that purely rational decisions maximize utility is a theoretical construct with notable limitations. Wierzbicki and Nakamori (2006) observe that a utility function can often be devised to justify almost any decision, underscoring the need to incorporate practical considerations and real-world constraints into the decision-making process.

Ultimately, effective decision-making requires a synthesis of logic and intuition. By embracing both, individuals can navigate complexity, make well-informed judgments, and achieve their goals more effectively. Building on this foundation, Chapter 5 will introduce a decision-making theory rooted in knowledge synthesis. But first, we will explore creative decision-making and its role in enriching this process.

4.1.2 Abduction (A Form of Hypothetical Reasoning)

Abduction, introduced by the pragmatist philosopher Charles Peirce, is a form of reasoning that generates hypotheses to explain unexplained phenomena. Also known as hypothesis generation or hypothetical reasoning, it represents a third mode of reasoning alongside induction and deduction, as recognized by Aristotle.

Peirce argued that abduction is essential to scientific thought, though it is not the sole driver of discovery. Observation, experimentation, and rigorous testing are equally fundamental to the scientific process.

Abductive reasoning follows this general pattern:

- We observe B.
- If we assume A, then B can be explained.
- Therefore, A is a possible explanation for B.

Although abductive reasoning does not yield definitive answers—since alternative explanations for B may exist—it is a valuable tool for generating new ideas and hypotheses that can be further tested and refined.

For example, consider the following case:

- Observation: The grass is wet.
- Hypothesis: It rained last night.
- Conclusion: Rainfall is a plausible explanation for the wet grass.

Abduction is a form of logical inference that begins with specific observations and works backward to propose a plausible explanation. Unlike deduction, which moves from general principles to particular conclusions, abduction formulates hypotheses that provide potential insights or discoveries. While it does not guarantee the truth of its conclusions, it often serves as a foundation for further inquiry. Table 4.1 compares three reasoning methods.

Table 4.1 Comparison of reasoning methods

Reasoning method	Process	Outcome	Information
Induction	Specific observations → general principle	Probable conclusion	Increases
Deduction	General principle → specific conclusion	Certain conclusion	Does not increase
Abduction	Specific observations → plausible explanation	Possible hypothesis	Increases

4.1.2.1 Creative Imagination

Logical reasoning, particularly induction and deduction, is widely used in business to gain a competitive edge. However, relying solely on these methods can lead to homogeneous thinking, limiting the exploration of novel ideas and solutions. To overcome this, abduction—also known as hypothetical reasoning—is gaining recognition for its ability to generate creative value.

As shown in Table 4.1, deduction and abduction differ significantly in approach. Deduction derives specific conclusions from general principles, moving from cause to effect. In contrast, abduction infers a plausible cause for a given effect. While induction also seeks to infer causes, it does so by identifying patterns from repeated observations.

Newton's work on gravity illustrates how these forms of reasoning interact. His observations of falling apples and other phenomena supported inductive generalizations about gravity. Meanwhile, his hypothesis that a universal gravitational force governs both terrestrial and celestial motion exemplifies abduction.

Abductive reasoning generates hypotheses that must be tested and refined through deduction and induction. Deduction predicts a hypothesis's implications, which are then validated or refuted through experimentation. When evidence consistently supports a hypothesis, induction generalizes these findings into a broader theory. Newton's universal law of gravitation emerged from this iterative interplay of abduction, deduction, and induction, demonstrating the power of integrating multiple reasoning methods.

To gain a competitive advantage, businesses must first leverage inspiration—drawing analogies, exploring intuitive insights, and employing techniques such as mind mapping and brainstorming to generate unique hypotheses from limited observations. However, true innovation and sustained success require rigorous validation and refinement of these hypotheses through deduction and induction.

This process is inherently iterative: Abduction generates novel hypotheses, deduction tests their implications, and induction generalizes findings into broader principles. By effectively integrating these three reasoning methods, businesses can foster creativity, drive innovation, and secure a lasting competitive edge.

4.1.2.2 Abductive Reasoning in Research

Although scientific research is often perceived as strictly logical, creative imagination frequently underpins its foundation. Research usually begins with observations that challenge existing theories, contradict established models, or reveal unexpected anomalies. Abductive reasoning enables researchers to creatively apply their knowledge and imagination, proposing explanations for these anomalies and opening new avenues of inquiry.

- *Natural Sciences*: Abduction is invaluable for interpreting unexpected experimental results or observations that conflict with established models. For example,

Alexander Fleming's discovery of penicillin began with a simple observation: the inhibition of bacterial growth around a mold contaminant. This led to the abductive hypothesis that the mold produced a substance capable of killing bacteria—an insight that ultimately paved the way for antibiotics.

- *Social Sciences*: Social scientists frequently employ abduction to interpret qualitative data, such as interviews or ethnographic observations. For instance, a sudden surge in support for a charismatic political candidate might lead to the hypothesis that voters are seeking a fresh approach to politics, reflecting broader societal dissatisfaction with the status quo.
- *Humanities*: In disciplines such as history and literature, abduction helps researchers interpret texts, artifacts, and cultural phenomena. Art historians, for example, might hypothesize about an artist's intentions, the social context of a work's creation, or the meaning embedded within it.

Abduction allows researchers to move beyond mere description, exploring underlying causes and mechanisms. By identifying plausible explanations for unexpected observations, abduction guides the formulation of research questions and study designs. However, to ensure validity and reliability, rigorously evaluating these hypotheses using complementary methods, such as experiments, surveys, or textual analysis, is essential.

4.1.2.3 Abductive Reasoning in Business

Abduction, which infers the most plausible explanation for an observation, is a valuable tool for businesses seeking innovation and growth. By generating and testing hypotheses, abduction informs strategic decision-making across various domains:

- Product Development: Identifies market trends (e.g., regional popularity) and hypothesizes underlying factors, such as cultural preferences or competitive advantages.
- Strategic Planning: Analyzes declining market share and hypothesizes potential causes, including competitor strengths or unmet customer needs.
- Problem-Solving: Investigates operational challenges, such as production delays due to machine malfunctions or material degradation.

By emphasizing hypothesis generation and testing, abduction provides significant advantages for businesses:

- Fostering Innovation: Encourages unconventional thinking and exploration of multiple perspectives, sparking novel ideas and solutions.
- Accelerating Problem-Solving: Facilitates rapid identification of root causes and effective solutions through structured hypothesis testing.
- Mitigating Risk: Supports informed decision-making by considering multiple possibilities, reducing uncertainty and potential risks.
- Enhancing Creativity: Promotes out-of-the-box thinking, leading to innovative approaches to business challenges.

Abduction is a powerful tool for problem-solving and value creation. It helps organizations cultivate a culture of innovation and drive sustainable growth. By intentionally applying abductive reasoning, businesses can make more adaptive, insightful, and forward-thinking decisions in an ever-evolving marketplace.

4.1.2.4 Menu Development Through Abduction

Abductive reasoning can play a pivotal role in menu planning for a conveyor belt sushi restaurant. By observing trends, analyzing customer data, and studying competitors, the restaurant can:

- Identify Customer Needs: Develop and test hypotheses about preferences (e.g., "Demand for vegetarian options is increasing").
- Gain a Competitive Edge: Differentiate itself with distinctive offerings, such as fusion dishes or expanded dessert selections.
- Utilize Seasonality: Incorporate seasonal ingredients (e.g., "Summer rolls with fresh vegetables") to create appealing limited-time specials.
- Capitalize on Culinary Trends: Attract diners with trendy dishes, such as visually striking, social media-friendly platters.

Beyond Prediction: Abduction is more than a predictive tool—it fuels creativity and innovation. By gaining deeper insights into customer preferences, the restaurant can craft a compelling menu that captivates diners, featuring dishes like:

- Vibrant Summer Rolls: Fresh summer vegetables wrapped in delicate rice paper.
- Refreshing Seafood Noodle Bowls: Light, flavorful dishes perfect for warm weather.
- Decadent Fruit Desserts: Sweet and indulgent desserts that surprise and delight.

By applying abductive reasoning, the restaurant can:

- Identify evolving customer preferences.
- Inspire innovative menu concepts.
- Enhance customer satisfaction.

Leveraging abduction enables the restaurant to craft a distinctive, trend-driven menu, setting itself apart from competitors and driving long-term success.

4.1.2.5 Abductive Reasoning: A Powerful Tool for Decision-Making

Abductive reasoning, often called "inference to the best explanation," is vital for effective decision-making. It helps individuals and organizations in several key ways:

1. Generating Hypotheses: Identifies patterns and anomalies in data, sparks potential explanations, and fosters innovative thinking by exploring diverse possibilities.

2. Evaluating Information: Prioritizes relevant data, reveals knowledge gaps, and guides further inquiry, ensuring that decisions are well-informed.
3. Making Informed Decisions: Weighs multiple possibilities, mitigates cognitive biases, and prevents premature conclusions, particularly in uncertain or ambiguous situations.
4. Learning and Adapting: Encourages continuous improvement through hypothesis testing and feedback loops, enabling individuals and organizations to refine strategies effectively.

Abductive reasoning empowers individuals to navigate uncertainty with creativity and critical thinking. By fostering innovative problem-solving, thorough evaluation, and adaptive learning, it serves as a cornerstone of personal and professional success.

4.2 Knowledge Creation and Management

Knowledge creation and management are essential for making informed decisions at both the organizational and individual levels. By facilitating access to relevant information and insights, these processes enable data-driven decision-making rather than reliance on assumptions or intuition. This section introduces a key knowledge creation and management model widely applied in organizational and corporate settings.

4.2.1 Organizational Knowledge Creation

Organizational knowledge creation refers to generating, acquiring, sharing, and applying knowledge to enhance performance and achieve strategic objectives. It leverages the collective intelligence of employees, who collaboratively refine and implement knowledge to drive success.

Key components of organizational knowledge creation include:

- *Knowledge Generation*: Developing new ideas and insights.
- *Knowledge Acquisition*: Capturing external knowledge and expertise.
- *Knowledge Sharing*: Disseminating knowledge across teams and individuals.
- *Knowledge Utilization*: Applying knowledge to solve problems and achieve objectives.

Effective knowledge management enhances innovation, improves decision-making, increases efficiency, and provides a sustainable competitive advantage. Organizations that foster a culture of continuous learning and knowledge sharing are better positioned for long-term success.

Various models have been developed to manage and create knowledge systematically. Among them, the SECI Model (Nonaka & Takeuchi, 1995) is one of the most influential, offering a structured framework for how knowledge is created, shared, and applied within organizations.

4.2.1.1 Tacit Knowledge: A Key Component of the SECI Model

At the heart of the SECI model is tacit knowledge that is difficult to articulate or express explicitly. Tacit knowledge is deeply personal, intuitive, and experience-based, making it essential for innovation and decision-making.

Examples of tacit knowledge include:

- Physical Skills: Riding a bicycle, playing a musical instrument, or mastering a craft.
- Cognitive Skills: Intuitive problem-solving, pattern recognition, and rapid decision-making.
- Social Skills: Effective communication, negotiation, and leadership.
- Cultural Understanding: An innate awareness of societal norms, values, and behaviors.

The concept of tacit knowledge was introduced by Michael Polanyi (1958, 1966), who emphasized its critical role in knowledge synthesis. In Chapter 5, we will further explore his ideas, particularly the concept of tacit integration and its implications for decision-making.

4.2.1.2 The SECI Model: A Framework for Knowledge Creation

The SECI model (Nonaka & Takeuchi, 1995) is a foundational framework for understanding and facilitating organizational knowledge creation. The acronym SECI represents four core processes: Socialization, Externalization, Combination, and Internalization, as illustrated in Fig. 4.1.

The SECI model proposes that new knowledge emerges through the continuous and dynamic interplay between tacit and explicit knowledge. This ongoing exchange leads to the development of increasingly complex and enriched knowledge. The model underscores the importance of a cyclical knowledge creation process, where tacit knowledge is shared, converted into explicit knowledge, and subsequently refined and internalized as new tacit knowledge.

The SECI model consists of four interrelated stages that drive knowledge creation:

1. *Socialization*: Tacit knowledge, which is personal and difficult to articulate, is shared through direct social interactions and shared experiences. Knowledge sharing occurs through observation, apprenticeships, or informal discussions.
2. *Externalization*: Tacit knowledge is converted into explicit knowledge through documentation, conceptual modeling, or diagram creation. This stage involves articulating and codifying personal insights for broader use.

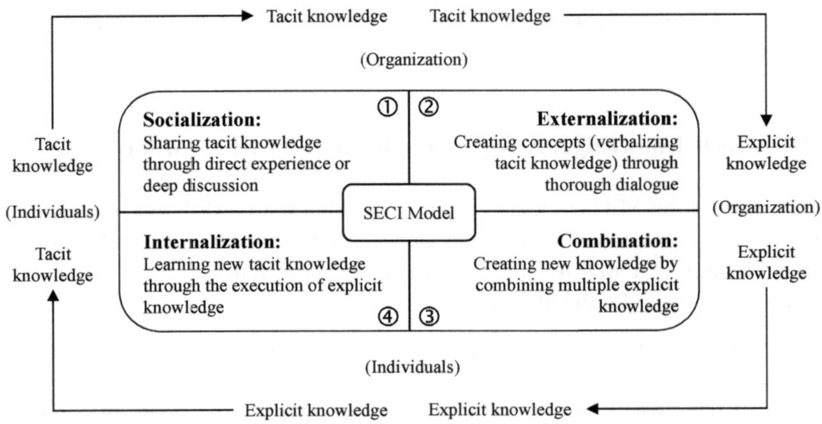

Fig. 4.1 The SECI model: a widely recognized framework for organizational knowledge creation

3. *Combination*: Explicit knowledge from various sources is synthesized and structured to generate new, actionable knowledge. This process involves integrating diverse perspectives, organizing information, or applying knowledge in novel ways.
4. *Internalization*: Newly created explicit knowledge is absorbed by individuals, becoming part of their tacit knowledge base through practice, reflection, or real-world application.

The SECI model highlights the crucial interplay between tacit and explicit knowledge in fostering organizational learning and innovation. It emphasizes the need for organizations to:

- Cultivate a knowledge-sharing culture.
- Encourage the transformation of knowledge into actionable insights.
- Apply knowledge effectively to drive innovation and maintain a competitive edge.

By fostering a structured approach to knowledge exchange, refinement, and application, the SECI model serves as a powerful tool for organizations aiming to sustain long-term success.

Applying the SECI Model—Challenges and Considerations: The SECI model provides a valuable framework for understanding knowledge creation, yet its practical implementation presents notable challenges. The following hypothetical example from software development illustrates key activities and associated difficulties within each stage:

1. *Socialization* (Tacit-to-Tacit)

   - *Activity*: Senior developers share experiences and best practices with junior team members through mentoring, pair programming, and code reviews.

- *Challenge*: Effective knowledge transfer requires a shared context and mutual understanding. Differences in background, experience, or communication styles may hinder the successful transmission of tacit knowledge.

2. *Externalization* (Tacit-to-Explicit)
   - *Activity*: Developers document design decisions, coding practices, and workflows in manuals, wikis, and code comments.
   - *Challenge*: Converting tacit knowledge into an explicit form is inherently complex. Capturing intuitive skills and problem-solving techniques in clear, concise documentation requires strong communication skills and a clear understanding of the audience's needs.
3. *Combination* (Explicit-to-Explicit)
   - *Activity*: The team integrates code libraries, frameworks, and design patterns to develop new software features.
   - *Challenge*: Effectively synthesizing diverse sources of explicit knowledge demands creativity, critical thinking, and a systems-level perspective to ensure seamless integration and prevent inefficiencies.
4. *Internalization* (Explicit-to-Tacit)
   - *Activity*: Developers gain hands-on experience with the software, internalizing its functionality and principles through practice and application.
   - *Challenge*: Successful internalization requires more than exposure to new knowledge. It necessitates active engagement, iterative practice, and contextual application to integrate new insights into individual mental models.

Recognizing these challenges, organizations can adopt targeted strategies to facilitate effective knowledge creation:

- Fostering collaboration through structured mentoring and peer learning.
- Providing training in documentation and effective communication.
- Encouraging experiential learning through hands-on projects and real-world application.

Addressing these obstacles will allow organizations to fully leverage the SECI model, fostering a culture of continuous learning and sustained innovation.

The SECI Model in Action—Summer Menu Innovation: The SECI model offers a structured approach to developing innovative summer menu ideas for a conveyor belt sushi restaurant. The following outlines how each stage of the SECI model can be applied to the menu development process:

1. *Socialization* (Tacit-to-Tacit): Team members exchange personal experiences, intuitions, and insights about summer dining preferences, popular dishes, and seasonal ingredients. This stage may involve discussions on past summer promotions, customer feedback, and emerging industry trends.

2. *Externalization* (Tacit-to-Explicit): Tacit knowledge is articulated and documented through brainstorming sessions, where ideas are generated, categorized, and refined. Key themes—such as "light and refreshing," "family-friendly," and "seasonal specialties"—emerge as guiding concepts for menu development.
3. *Combination* (Explicit-to-Explicit): The team integrates explicit knowledge to develop innovative menu concepts. For example, merging the themes "light and refreshing" and "seasonal specialties" could inspire a summer roll featuring fresh vegetables, fragrant herbs, and a tangy dipping sauce.
4. *Internalization* (Explicit-to-Tacit): Team members internalize the new menu concepts through hands-on practice, such as test kitchen trials and customer feedback sessions. This experiential process helps refine their understanding of ingredient pairings, preparation techniques, and customer preferences.

Taking these insights into account, the restaurant can develop the following summer menu offerings:

- Light and Refreshing: Cold soba noodles with seafood, summer rolls with fresh vegetables and tangy dipping sauce, and chilled seafood salads
- Family-Friendly: Colorful sushi rolls shaped like cute animals, kids' meal sets with playful side dishes, and dessert sushi featuring seasonal fruits
- Seasonal Specialties: Sushi featuring seasonal fish like mackerel or sea bream, cold soup dishes with refreshing flavors, and limited-time summer rolls with unique fillings

By applying the SECI model, the restaurant can develop an innovative summer menu to captivate customers, strengthen brand engagement, and boost sales.

4.2.1.3 Key Benefits of the SECI Model for Decision-Making

The SECI model offers a robust framework for enhancing organizational decision-making by integrating tacit and explicit knowledge. Its application delivers multiple advantages across key areas. Below are some of the most significant benefits organizations can experience:

1. Leveraging Tacit Knowledge: Organizations can tap into the expertise of veteran employees, drawing on their valuable insights and experiences. Integrating tacit knowledge leads to more informed, well-rounded, and innovative decisions.
2. Stimulating Creativity: The SECI model fosters creativity by blending tacit and explicit knowledge to generate new ideas and solutions. Using past experiences and lessons learned as a foundation, organizations can develop innovative strategies while reducing risks associated with new initiatives.
3. Promoting Knowledge Transfer: The SECI model facilitates knowledge transfer across generations of employees, ensuring that critical insights and skills are preserved. This continuous exchange enhances organizational learning and drives ongoing improvement.

Implementing the SECI model helps organizations build a culture of collaboration, creativity, and learning, equipping them to adapt to challenges and capitalize on new opportunities effectively.

4.2.2 Knowledge Management for Decision-Making

Knowledge management has evolved significantly since its emergence in the 1980s. Initially driven by the rise of computers in business, early efforts focused on collecting and sharing practical organizational knowledge. The Internet's rapid expansion in the 1990s further propelled the field, with Ikujiro Nonaka's SECI model providing a structured framework for knowledge creation and transfer. More recently, advancements in artificial intelligence (AI) and big data have revolutionized methodologies, enhancing how organizations manage, analyze, and apply knowledge.

Several key factors highlight the growing importance of knowledge management:

- Information Overload: The digital era has generated vast amounts of data, making it challenging to locate and leverage relevant information.
- Knowledge as a Competitive Advantage: In today's knowledge-driven economy, intellectual capital—such as employee expertise and customer insights—is a key differentiator.
- Knowledge Retention Challenges: Losing institutional knowledge due to employee turnover or retirement threatens organizational continuity.
- Customer-Centric Strategies: Addressing evolving customer needs requires access to and effectively applying customer-related knowledge.

Effective knowledge management relies on a comprehensive, multi-faceted approach that includes:

- Knowledge Identification: Recognizing and prioritizing critical knowledge assets.
- Knowledge Capture: Gathering and preserving knowledge in various formats, such as documents, databases, and expert interviews.
- Knowledge Organization: Structuring and categorizing knowledge for easy retrieval and accessibility.
- Knowledge Sharing: Facilitating dissemination through intranets, repositories, and collaboration platforms.
- Knowledge Application: Leveraging knowledge to enhance decision-making, drive innovation, and optimize business processes.

By effectively managing knowledge, organizations can navigate complex challenges, maintain a competitive edge, and adapt to evolving market demands.

4.2.2.1 Enhancing Decision-Making Through Knowledge Management

Effective decision-making depends on the availability and application of relevant knowledge. Knowledge management (KM) provides a robust framework to support and enhance this process. Below are key techniques demonstrating how KM drives better decision-making:

- Knowledge-Based Decision Support Systems: These systems integrate KM principles with decision support tools to deliver actionable insights. By capturing and organizing expert knowledge, they help decision-makers analyze complex problems, identify solutions, and evaluate alternatives more effectively.
- Scenario Planning with Knowledge Bases: Scenario planning envisions future states and outcomes. Knowledge bases enhance this process by providing access to historical data, expert insights, and lessons learned. They enable decision-makers to identify key uncertainties, develop alternative scenarios, and assess potential impacts.
- Collaborative Decision-Making with Knowledge Sharing: Many decisions require input from diverse stakeholders. KM fosters collaboration by establishing platforms for seamless information exchange, allowing teams to share insights and arrive at well-informed, robust decisions.
- Root Cause Analysis Through Knowledge Retrieval: Identifying the root cause of a problem is crucial for effective decision-making. Knowledge retrieval systems provide access to past experiences, case studies, and lessons learned, helping organizations uncover underlying issues and devise practical solutions quickly.
- Risk Assessment with Knowledge-Based Expert Systems: Assessing and mitigating risks is vital for organizational success. KM, combined with expert systems, supports risk management by analyzing risk factors, evaluating potential impacts, and recommending mitigation strategies based on historical data, expert opinions, and best practices.
- Knowledge-Driven Innovation: Innovation thrives when organizations effectively combine existing knowledge with new ideas. KM nurtures innovation by capturing, sharing, and leveraging knowledge, paving the way for new products, services, and business models.

These techniques illustrate how KM enhances decision-making across all organizational levels. Their application varies based on the nature of the decisions, organizational needs, and available technology.

4.2.2.2 Benefits and Challenges of Knowledge Management

Organizations have gained significant advantages from effective knowledge management, as summarized in Table 4.2. Once considered a "nice-to-have," knowledge management is now a strategic imperative for organizations seeking to thrive in today's dynamic and competitive business environment. By capturing, sharing, and utilizing knowledge effectively, organizations can drive innovation, enhance decision-making, and foster a culture of continuous learning.

While knowledge management offers substantial benefits, it also presents challenges, as outlined in Table 4.3. Overcoming these obstacles requires a multi-faceted approach, including fostering a knowledge-sharing culture, breaking down organizational silos, providing adequate training and support, and leveraging technology effectively.

Table 4.2 Benefits of knowledge management

Benefit	Description
Improved decision-making	Access to relevant knowledge that enables more informed and strategic decisions
Fostering innovation	A culture of creativity and idea generation supported by knowledge sharing and best practices
Increased efficiency	Streamlined processes and reduced redundancy through easy access to knowledge
Knowledge retention	Prevention of critical knowledge loss due to employee turnover or retirement
Competitive edge	Strategic use of knowledge as a key asset to differentiate the organization in the marketplace

Table 4.3 Challenges of knowledge management

Challenge	Description
Resistance to change	Employees may resist sharing knowledge due to job security concerns, effort involved, or cultural barriers
Knowledge silos	Departments retain information excessively, preventing knowledge from flowing freely, leading to inefficiency
Information overload	The overwhelming volume of data makes finding and utilizing relevant information challenging
Capturing tacit knowledge	Tacit knowledge is personal and experience-based, making it difficult to document; techniques like mentoring help
Maintaining knowledge quality	Ensuring the accuracy and relevance of stored knowledge is an ongoing challenge to prevent poor decisions
Measuring return on investment	Demonstrating the return on investment for knowledge management requires clear performance indicators and tracking mechanisms
Technology challenges	Implementing and maintaining systems can be costly and complex, with integration issues often arising
Lack of leadership support	Knowledge management initiatives are unlikely to succeed without senior management support and commitment

4.2.2.3 Knowledge Management in the Age of AI

The convergence of artificial intelligence (AI) and knowledge management (KM) transforms how organizations capture, store, and utilize information. This powerful combination unlocks new possibilities:

1. Enhanced Knowledge Capture and Organization
    - AI-powered natural language processing (NLP) extracts insights from unstructured data (e.g., customer service transcripts and social media) with unprecedented speed and accuracy.
    - Semantic search goes beyond keywords by understanding context and meaning, delivering highly relevant results.
    - Automation streamlines data entry, classification, and tagging, freeing up human resources for higher-value tasks.
2. Improved Knowledge Sharing and Collaboration
    - Personalized recommendations deliver tailored content to users, maximizing relevance and engagement.
    - AI-powered assistants provide instant access to information, answer queries, and guide users through knowledge repositories.
    - Enhanced collaboration tools, such as real-time translation, automatic summarization, and intelligent meeting assistants, facilitate seamless communication and teamwork.
3. Advanced Knowledge Analysis and Insights
    - Predictive analytics identifies patterns and trends, enabling proactive decision-making and risk mitigation.
    - Knowledge graphs visualize complex relationships between concepts, offering a holistic understanding of the organization's knowledge landscape.
    - Continuous learning refines AI models over time, ensuring increasingly accurate and valuable insights.
4. Ethical Considerations and Bias Mitigation
    - Data diversity is essential to prevent AI from perpetuating biases.
    - Transparency and explainability are critical for building trust and ensuring ethical decision-making.
5. Integration with Emerging Technologies
    - IoT (Internet of Things) integration leverages real-time data from connected devices to optimize operations and enhance decision-making.
    - Blockchain technology provides a secure and immutable foundation for storing and sharing sensitive knowledge.

4.2.2.4 The Future of Knowledge Management

AI is revolutionizing knowledge management (KM) by automating routine tasks, enhancing search capabilities, and extracting deeper insights from data. However, human expertise remains indispensable. By combining human ingenuity with AI, organizations can:

- Drive innovation by fostering a culture of continuous learning and knowledge sharing.
- Gain a competitive advantage by making data-driven decisions with incredible speed and accuracy.
- Enhance customer experience through faster, more personalized service.

To effectively capitalize on these opportunities, organizations should focus on the following key strategies:

- Embrace AI-powered KM solutions while maintaining ethical considerations and human oversight.
- Invest in data quality and diversity to ensure accurate and unbiased AI-driven insights.
- Cultivate a culture of collaboration to promote knowledge sharing across the organization.

By strategically integrating AI with a human-centered approach, organizations can unlock the full potential of KM and thrive in the digital age.

4.3 Idea Creation Frameworks

This section explores structured frameworks for idea generation, within and beyond the business world. These methodologies enhance problem-solving, decision-making, and strategic planning by providing systematic approaches to tackling complex challenges.

Business frameworks are structured models that help analyze intricate problems and make well-informed decisions. By breaking down complexity into manageable components, they enable deeper understanding and more effective strategy development.

Key benefits of business frameworks include:

- Clarity and Focus: Help clarify thinking and identify key issues.
- Informed Decision-Making: Provide a structured approach that reduces bias and improves outcomes.
- Effective Problem-Solving: Enable analysis from multiple perspectives to develop practical solutions.
- Strategic Planning: Support the creation and execution of actionable strategies to achieve long-term goals.

4.3.1 Business Frameworks

This subsection introduces several widely used business frameworks, which serve various purposes, such as:

A. *Identifying Issues*: Identifying specific business problems or challenges.
B. *Discovering Needs*: Uncovering unmet customer or market needs.
C. *Generating Ideas*: Brainstorming innovative solutions and approaches.
D. *Formulating Strategies*: Developing comprehensive plans to achieve specific goals.
E. *Improving Operations*: Enhancing efficiency and effectiveness in business processes.

These categories often overlap, as many frameworks can be applied to multiple purposes. A combined approach can provide a more comprehensive understanding of a problem.

Many of the frameworks below originate primarily from management consulting firms, quality improvement methodologies, or general problem-solving approaches, making them difficult to attribute to specific individuals or publications.

4.3.1.1 Identifying Issues

1. *AS-IS/TO-BE* (A Framework for Transformation): This framework compares the current state (AS-IS) of a process or system with its desired future state (TO-BE). By systematically analyzing the differences, organizations or individuals can identify key gaps, prioritize areas for improvement, and develop targeted action plans to achieve their desired outcomes.
2. *5W1H* (A Framework for Problem-Solving and Decision-Making): This is a simple yet effective framework for problem-solving and decision-making. By systematically addressing the following questions, one can gain a deeper understanding of the situation and develop practical solutions:
    - What: Clearly define the problem or goal. What is the specific issue or desired outcome?
    - Why: Identify the root causes of the problem or the reasons for pursuing the goal. Why is it important to address this issue?
    - Who: Determine who is involved in the problem or goal. Who are the stakeholders, both internal and external?
    - When: Establish a timeline for addressing the problem or achieving the goal. When should the solution be implemented and the desired outcome be realized?
    - Where: Identify the location or context in which the problem occurs or the goal needs to be achieved. Where is the problem manifesting, and where should the solution be applied?

- How: Develop strategies and tactics to address the problem or achieve the goal. How can the problem be solved or the goal be accomplished?

3. *6W2H* (A Comprehensive Framework for Problem-Solving and Decision-Making): Building upon the 5W1H framework, 6W2H introduces two additional questions to refine analysis further:

   - Whom: Identify the target audience or beneficiaries of the solution. Who will be directly impacted by it?
   - How Much: Assess the resources required to implement the solution, including time, budget, and personnel. How much will it cost to address the problem or achieve the goal?

4. *Logic Tree* (A Visual Tool for Problem-Solving and Decision-Making): A logic tree is a visual representation that breaks down a complex problem or goal into smaller, more manageable components. By systematically breaking down a problem into its root causes or mapping out the steps needed to achieve a goal, a logic tree provides a structured visualization that clarifies thinking, highlights key issues, and facilitates informed decision-making. This framework is particularly useful for tackling complex problems. We will explore this in greater detail in Sect. 4.3.2.1.
5. *Business Ethnography* (A Deep Dive into Organizational Culture): Business ethnography (Moeran, 2005; Spradley, 1980) is a qualitative research method in which researchers immerse themselves in an organization or cultural setting to gain a deeper understanding of its behaviors, values, and beliefs. This method typically involves direct interaction with participants through interviews, observations, and document analysis, making it particularly valuable for gaining insights into consumer behavior, organizational culture, and innovation. Key characteristics of business ethnography include:

   - Immersion: Researchers spend extended periods within the business environment, observing and participating in daily activities.
   - Observation: Researchers carefully observe individuals' behavior, interactions, and the physical and social context of their work.
   - Interviews: Researchers conduct in-depth interviews with employees, customers, and other stakeholders to gather qualitative data.
   - Document Analysis: Researchers analyze relevant documents, such as company reports, emails, and memos, to gain a deeper understanding of the organization.
   - Participant Observation: Researchers actively participate in the activities of an organization to gain firsthand experience and insights.

4.3.1.2 Discovering Needs

1. *PEST Analysis* (A Framework for Macro-Environmental Analysis): PEST analysis (Aguilar, 1967; Johnson et al., 2008) examines macro-environmental factors

that may significantly impact an organization. To gain a better understanding of the external environment and make informed decisions, PEST systematically analyzes the following four factors:

- Political Factors: Government policies, political stability, tax policies, trade regulations, and other government interventions that can affect a business.
- Economic Factors: Economic growth, interest rates, exchange rates, inflation rates, disposable income, and other indicators impacting business performance.
- Sociocultural Factors: Cultural norms and values, demographics, lifestyle changes, health consciousness, and other social and cultural trends influencing consumer behavior and market demand.
- Technological Factors: Technological advancements, the rate of technological change, and the level of innovation, which can disrupt industries and create new opportunities.

2. *Porter's Five Forces Analysis* (A Framework for Industry Analysis): Porter's five forces analysis (Porter, 1979) is a framework used to analyze an industry's competitive intensity and attractiveness. The five forces are:

   - Threat of New Entrants: Assesses how easily new competitors can enter the market. Factors influencing this include barriers to entry, such as high capital requirements, strong brand loyalty, and government regulations.
   - Bargaining Power of Suppliers: Evaluates the power of suppliers to influence prices and quality. Suppliers with strong bargaining power can demand higher prices, reduce quality, or limit supply.
   - Bargaining Power of Buyers: Assesses buyers' ability to influence prices and quality. Buyers with strong bargaining power can demand lower prices, higher quality, or additional services.
   - Threat of Substitute Products or Services: Measures the availability of alternative products or services that satisfy customer needs. The presence of substitutes can limit an industry's pricing power and profitability.
   - Competitive Rivalry: Examines the intensity of competition among existing firms. High rivalry can lead to price wars, aggressive marketing campaigns, and product innovations.

3. *SWOT Analysis* (A Strategic Planning Tool): SWOT analysis (Learned et al., 1965) is a strategic planning framework that helps individuals and organizations identify internal strengths and weaknesses as well as external opportunities and threats. We will delve into the SWOT analysis in Sect. 4.3.2.2.
4. *Empathy Map* (A Collaborative Tool for User Understanding): An empathy map (Gray et al., 2010) is a collaborative tool that helps teams better understand their users by visualizing their thoughts, feelings, motivations, and behaviors. It encourages teams to step into their users' shoes and consider:

   - What they say: Statements users make about their needs and challenges.

- What they think: Their internal thoughts and concerns, which may not always be expressed.
- What they feel: Emotional responses to a product or service.
- What they do: Observable actions and behaviors.
- Their pain points: Frustrations or difficulties they experience.
- Their desired gains: Aspirations and expectations for an improved experience.

5. *Customer Journey Map* (A Visual Representation of the Customer Experience): A customer journey map (Blokdyk, 2020) visually represents the customer experience, from initial brand awareness to post-purchase interactions. It helps identify touchpoints, pain points, and opportunities for improvement across various stages of the journey (e.g., awareness, consideration, purchase, retention). By understanding the customer's perspective, teams can optimize the customer experience and increase customer satisfaction.
6. *Value Chain Analysis* (A Framework for Strategic Analysis): Value chain analysis (Porter, 1985; Reese et al., 2016) is a strategic management tool that examines the activities involved in creating and delivering a product or service. By understanding the value chain, organizations can identify opportunities for improvement, cost reduction, and differentiation. The value chain is typically divided into two categories of activities:

   - Primary Activities (directly related to creating and delivering a product or service):
     - Inbound Logistics: Receiving, storing, and handling raw materials.
     - Operations: Transforming inputs into outputs, such as manufacturing or assembly.
     - Outbound Logistics: Storing and distributing finished products.
     - Marketing and Sales: Promoting and selling products or services.
     - Service: Providing customer support and maintenance.
   - Support Activities (supporting the primary activities):
     - Procurement: Sourcing raw materials and supplies.
     - Technology Development: Research and development of new technologies.
     - Human Resource Management: Recruiting, hiring, and training employees.
     - Infrastructure: Organizational systems and processes, such as finance and accounting.

7. *3C Analysis* (A Strategic Framework for Competitive Analysis): 3C analysis (Ohmae, 1982) is a strategic management framework used to analyze three key factors that influence a company's competitive position:

   - Company: Internal strengths, weaknesses, and core competencies.
   - Customer: The target customers' needs, wants, and behaviors.
   - Competitor: The competitive landscape, including the strengths, weaknesses, and strategies of competitors.

8. *4P Marketing Mix* (A Classic Framework for Marketing Strategy): The 4P marketing mix (Kotler et al., 2023; McCarthy, 1960) is a foundational framework introduced by E. Jerome McCarthy and later expanded by Philip Kotler and others. It focuses on four key elements:
   - Product: The company's goods or services, including their features, benefits, and branding.
   - Price: The pricing strategy, considering cost, competition, and perceived value.
   - Place: The distribution channels used to reach customers, including physical stores, online platforms, and other channels.
   - Promotion: The marketing and advertising strategies used to promote products or services, such as advertising, public relations, social media, and sales promotions.

4.3.1.3 Generating Ideas

1. *Brainstorming* (A Creative Technique for Idea Generation): Brainstorming (Osborn, 1953) is a creative technique for generating ideas. A group of people gathers to share ideas freely, without judgment. The goal is to generate as many ideas as possible, no matter how outlandish they may seem. Section 4.4.2.2 provides a concrete example of this technique.
2. *Mind Mapping* (A Visual Tool for Organizing Thoughts): Mind mapping (Buzan & Buzan, 1996) is a visual technique for organizing information and generating ideas. It involves creating a diagram with a central idea in the middle, surrounded by related ideas branching outward. Mind mapping can stimulate creativity, improve memory, and enhance problem-solving skills. Section 4.4.2.2 provides a concrete example of this technique.
3. *Affinity Diagram* (A Tool for Grouping and Categorizing Ideas): An affinity diagram (Kawakita, 1967) is a tool for organizing and categorizing large sets of ideas. Ideas are written on sticky notes and then grouped based on similarities. This technique helps identify patterns, themes, and potential solutions. Section 4.4.2.2 provides a concrete example of this technique.
4. *SCAMPER* (A Creative Problem-Solving Technique): SCAMPER (Eberle, 2008) is a creative problem-solving technique that encourages innovative thinking by systematically modifying an existing product or process. The acronym stands for:
   - Substitute: Replace a component or element with something different.
   - Combine: Merge two or more ideas, products, or processes.
   - Adapt: Adjust an idea or product to fit a new context or purpose.
   - Modify: Alter the form, function, or appearance of something.
   - Put to Another Use: Find new applications for an existing idea or product.
   - Eliminate: Remove unnecessary components or features.
   - Reverse: Invert the order, direction, or perspective of something.

Applying the SCAMPER method to an existing idea or product can generate fresh and innovative solutions. Osborn's Checklist, a broader brainstorming technique introduced by Osborn (1953), includes SCAMPER as a more structured and focused approach to creative problem-solving.

5. *Six Thinking Hats* (A Creative Thinking Technique): The six thinking hats method (de Bono, 1985) is a structured creative thinking technique that guides individuals in analyzing problems from six distinct perspectives, represented by different colored hats:

   - White Hat: Gathers data, facts, and neutral information to build an objective view.
   - Red Hat: Considers emotional reactions and instinctive responses without justification.
   - Black Hat: Identifies possible risks, challenges, and weaknesses to avoid pitfalls.
   - Yellow Hat: Highlights strengths, advantages, and potential benefits of an idea.
   - Green Hat: Encourages out-of-the-box thinking, innovation, and alternative approaches.
   - Blue Hat: Manages the thought process, sets objectives, and guides discussion flow.

4.3.1.4 Formulating Strategies

1. *Cross-SWOT Analysis* (An Advanced Strategic Planning Tool): This strategic analysis technique, popularized by Weihrich (1982), enhances the traditional SWOT analysis (Strengths, Weaknesses, Opportunities, Threats) by identifying strategic opportunities and threats through cross-matching internal and external factors. By cross-matching the internal factors (strengths and weaknesses) with the external factors (opportunities and threats), organizations can identify areas where strengths can be leveraged to exploit opportunities or where weaknesses need to be addressed to defend against threats. This approach enables organizations to develop strategies that:

   - Leverage strengths to maximize opportunities.
   - Mitigate weaknesses to minimize threats.
   - Capitalize on opportunities to drive growth.
   - Avoid threats to ensure sustainability.

   We will delve deeper into the Cross-SWOT analysis in Section 4.3.2.3.

2. *STP Analysis* (A Marketing Strategy Framework): STP analysis (Kotler et al., 2021) is a strategic marketing tool that enables companies to effectively segment their market, target the most promising segments, and position their brand or products in a way that resonates with consumers. The process involves three key steps:

- Segmentation: Dividing the market into distinct groups of consumers with similar needs, wants, or behaviors.
- Targeting: Selecting one or more segments to focus on based on their potential profitability and alignment with the company's capabilities.
- Positioning: Creating a unique and compelling brand image for the target market to differentiate the product or service from competitors.

3. *Positioning Map* (A Strategic Tool for Market Positioning): A positioning map (Aaker, 2003; Kotler et al., 2023; Ries & Trout, 1981), also known as a Perceptual Map, is a visual tool for comparing brands or products based on two or more dimensions, such as price vs. quality or innovation vs. tradition. This tool helps marketers visualize their brand's positioning relative to competitors and identify opportunities to either reinforce their brand's unique qualities or adjust strategies to compete more effectively. The process of creating and analyzing a positioning map involves the following steps:

   - Identify Key Attributes: Determine the two most important attributes or dimensions that differentiate products or brands in the market.
   - Create Axes: Plot these two attributes on a graph, with one attribute on the x-axis and the other on the y-axis.
   - Position Brands: Place each brand or product on the map based on its perceived position on the two attributes.
   - Analyze the Map: Identify gaps in the market, areas of overcrowding, and potential differentiation opportunities.

4. *Business Model Canvas* (A Visual Strategy Framework): The business model canvas (Osterwalder & Pigneur, 2010) is a strategic tool that helps organizations map out the key components of their business model, including customer segments, value propositions, and revenue streams. It provides a visual representation of a company's value proposition, infrastructure, customer relationships, and financials. The nine building blocks of a business model canvas are:

   - Customer Segments: The different groups of people or organizations a company aims to serve.
   - Value Propositions: The unique value a company delivers to its customers.
   - Channels: The methods through which a company delivers its value propositions.
   - Customer Relationships: The types of relationships a company establishes with its customers.
   - Revenue Streams: The sources of revenue generated from each customer segment.
   - Key Resources: The most critical assets required to deliver the value proposition.
   - Key Activities: The essential tasks a company must perform to deliver its value proposition.
   - Key Partnerships: The network of partners and suppliers that support value delivery.

- Cost Structure: The most significant costs incurred by the company.

4.3.1.5 Improving Operations

1. *Six Sigma* (A Data-Driven Approach to Process Improvement): Six Sigma (Harry & Schroeder, 2000; Pande et al., 2001) is a set of tools and techniques aimed at enhancing the quality of products and services by identifying and eliminating defects and variations within processes. The term "Six Sigma" originates from Motorola's goal to reduce errors to just 3.4 occurrences per million operations, corresponding to 6σ. The core principles of Six Sigma include:
   - Customer Focus: Understanding and fulfilling customer needs and expectations.
   - Process Improvement: Identifying and removing defects and inefficiencies within processes.
   - Data-Driven Decision-Making: Relying on data to analyze problems and make informed decisions.
   - Continuous Improvement: Consistently working toward improvement in all areas of the business.
2. *PDCA Cycle* (A Cycle of Continuous Improvement): The PDCA Cycle, also known as the Deming Cycle or Shewhart Cycle, is a four-step iterative process for continuous improvement. Initially developed by Walter A. Shewhart in the 1920s and later popularized by W. Edwards Deming (Deming, 1982), it follows a cyclical approach to drive sustained results. The four steps of the PDCA Cycle are:
   - *Plan*: Define the problem or opportunity, set clear, measurable goals, and develop a detailed plan to implement change.
   - *Do*: Implement the plan on a small scale (pilot test) to gather data and observe the results.
   - *Check*: Analyze the data from the Do phase to evaluate the plan's effectiveness. Determine whether the desired outcomes were achieved and identify any issues.
   - *Act*: Standardize successful changes to incorporate them into regular processes. Take corrective actions for any problems encountered. Learn from the entire process and apply the insights to future improvements.

   The PDCA Cycle emphasizes continuous learning and improvement, encouraging organizations to assess and refine their processes for ongoing success regularly.

4.3.2 Decision-Making Frameworks

This subsection explores key frameworks that facilitate decision-making in various contexts, including non-business settings:

- *Logic Trees*: A hierarchical tool that breaks down complex problems into smaller, more manageable components. It helps identify root causes, potential solutions, and associated risks.
- *SWOT and Cross-SWOT Analysis*: SWOT analysis evaluates an organization's strengths, weaknesses, opportunities, and threats. Cross-SWOT analysis further examines interactions between these factors to uncover strategic opportunities

4.3.2.1 Logic Tree

A logic tree is a visual tool used to deconstruct complex issues into structured, hierarchical components. It enhances problem analysis, root cause identification, and solution development. There are three main types of logic trees:

- *What Tree* (Element Decomposition Tree): Breaks down a complex system or concept into its fundamental components, starting with a main topic and branching out into sub-elements.
- *Why Tree* (Cause Pursuit Tree): Traces the root causes of a problem, beginning with a problem statement and branching out into underlying reasons.
- *How Tree* (Problem-Solving Tree): Identifies solutions or strategies to address a problem by starting with a desired outcome and mapping out the necessary steps to achieve it.

Using logic trees can improve problem-solving skills, enhance critical thinking, and support more structured decision-making. Figures 4.2, 4.3, and 4.4 provide simple illustrations of these tree types, meant to demonstrate their structure rather than represent realistic scenarios.

Limitations: While logic trees are effective for breaking down problems into structured components, they may not be suitable for all situations. For example, using a logic tree to trace the cause of a disease from its symptoms can be challenging. Diseases often stem from a complex interplay of factors—genetics, environment, and lifestyle—making it difficult to pinpoint a single root cause. In such cases, a logic tree may oversimplify the issue, limiting its usefulness.

Key Principles for Creating an Effective Logic Tree: Building a robust logic tree requires careful attention to several key aspects:

1. Define a Clear Objective.
    - Purpose-Driven Approach: Begin by specifying the goal of your logic tree. Are you trying to:
        - Understand a complex system? (What Tree)

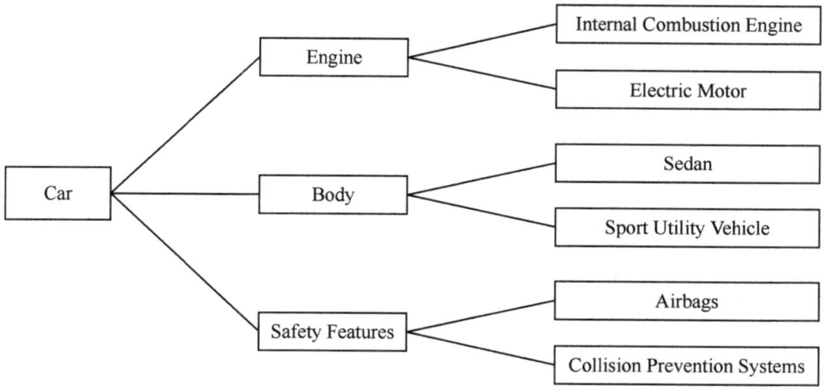

Fig. 4.2 What tree: breaking down the concept of a car

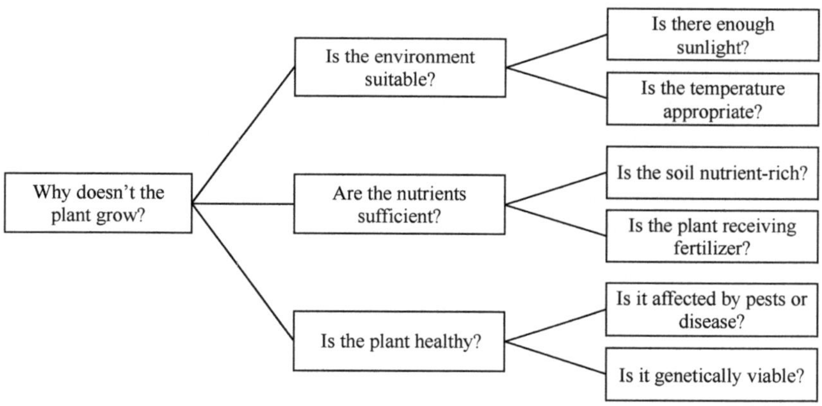

Fig. 4.3 Why tree: finding out why the plant doesn't grow

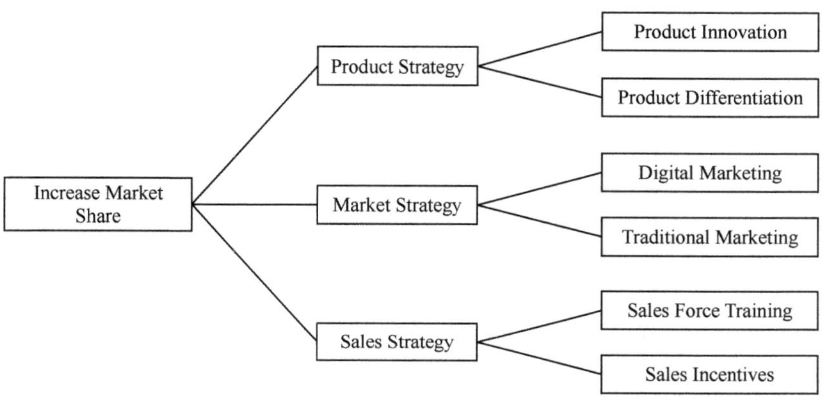

Fig. 4.4 How tree: inventing ways to increase market share

  - Identify root causes of a problem? (Why Tree)
  - Develop effective solutions? (How Tree)

- Select the Right Type: Choose the appropriate logic tree (What, Why, or How) based on your objective.

2. Follow the MECE Principle.

- Mutually Exclusive, Collectively Exhaustive (MECE): Ensure that all branches and sub-branches adhere to these criteria:

  - Mutually Exclusive: Each branch should be distinct, with no overlap.
  - Collectively Exhaustive: Together, all branches should comprehensively cover the issue.

- Systematic Analysis: Applying MECE ensures a structured, thorough approach to problem-solving.

3. Drill Down to Actionable Insights.

- Break Down into Manageable Components: Continue refining each branch until you reach a level of detail that yields actionable insights.
- Focus on Root Causes (Why Trees): Go deeper to uncover the fundamental causes of a problem.
- Prioritize Solutions (How Trees): Evaluate potential solutions based on impact, feasibility, and cost-effectiveness.

By following these principles, you can create effective logic trees that provide structured, insightful analyses, enabling better decision-making and problem-solving.

Summer Menu Planning for a Conveyor Belt Sushi Restaurant: Fig. 4.5 presents a logic tree designed to help a conveyor belt sushi restaurant plan its summer menu. By breaking down the process into manageable components, the logic tree helps identify potential menu items, evaluate their feasibility, and ultimately create a diverse and appealing seasonal menu.

4.3.2.2 SWOT Analysis

SWOT Analysis is a strategic planning technique that helps organizations assess their internal capabilities and external environment to develop effective strategies. This framework identifies an organization's Strengths, Weaknesses, Opportunities, and Threats:

- *Strengths*: Internal attributes that provide an advantage.
- *Weaknesses*: Internal limitations that hinder performance.
- *Opportunities*: External factors that the organization can leverage.
- *Threats*: External factors that pose risks to the organization.

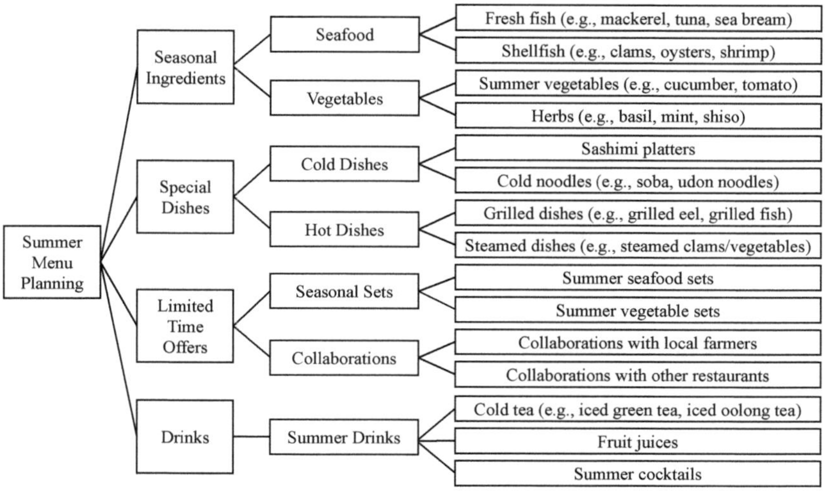

Fig. 4.5 A logic tree for planning a summer menu at a conveyor belt sushi restaurant

Understanding these four elements can help organizations capitalize on their strengths, address weaknesses, seize opportunities, and mitigate threats. While the exact origins of SWOT analysis are unclear, it is widely used today.

Figure 4.6 provides a template to facilitate SWOT analysis.

Internal Environment Analysis: Understanding an organization's internal environment is essential for identifying strengths and weaknesses that can impact its performance. Key internal factors include:

- Human Resources: Employee skills, experience, and motivation.
- Physical Resources: Tangible assets such as facilities, equipment, and inventory.

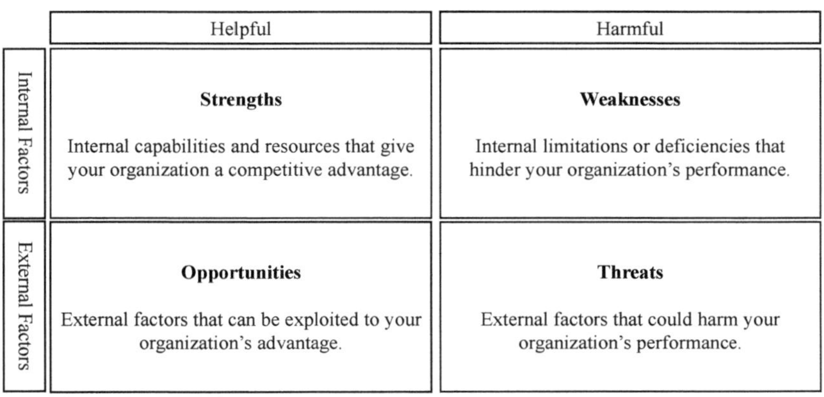

Fig. 4.6 SWOT analysis template

- Financial Resources: The organization's financial health, including revenue, expenses, and cash flow.
- Technological Resources: Infrastructure and technological capabilities.
- Brand Reputation: Public perception and brand image.

Organizations can leverage their strengths and address weaknesses to enhance performance:

A. *Strengths Analysis*: When evaluating strengths, consider:

   - Core Competencies: Unique skills or capabilities that provide a competitive edge.
   - Competitive Advantages: Differentiating factors that set the organization apart.
   - Strong Brand Reputation: Positive public image and customer loyalty.
   - Financial Stability: A solid financial position with sufficient resources.
   - Strong Leadership: Effective management and clear strategic direction.

B. *Weaknesses Analysis*: When assessing weaknesses, consider:

   - Lack of Expertise: Gaps in skills or knowledge.
   - Financial Constraints: Limited resources for growth and innovation.
   - Inefficient Processes: Outdated or ineffective operations that hinder productivity.
   - Negative Brand Perception: Poor reputation or declining customer trust.
   - Weak Leadership: Ineffective management and lack of strategic direction.

External Environment Analysis: A comprehensive understanding of the external environment is essential for organizations to identify opportunities and threats that may impact their operations and strategic direction. Key external factors include:

- Economic Factors: Economic conditions such as interest rates, inflation, and GDP growth can significantly influence organizational performance.
- Technological Factors: Advancements in technology can create new opportunities while disrupting existing markets.
- Sociocultural Factors: Shifts in societal values, attitudes, and lifestyles affect consumer behavior and preferences.
- Political Factors: Government policies, regulations, and political stability impact business operations.
- Legal Factors: Laws and regulations influence business practices, compliance requirements, and operational costs.
- Environmental Factors: Sustainability concerns and environmental regulations affect operations and product development.

Organizations can gain deeper insights into their external environment and make informed strategic decisions by analyzing opportunities and threats:

A. *Opportunity Analysis*: Opportunity analysis involves identifying and evaluating external factors that can drive growth, such as:

- Market Trends: Emerging consumer preferences and industry shifts.
- Competitive Landscape: Assessing competitors' strengths, weaknesses, and strategies.
- Technological Advancements: Leveraging innovations to enhance products or services.
- Regulatory Changes: Capitalizing on favorable policy adjustments.

B. *Threat Analysis*: Threat analysis focuses on identifying external risks that may challenge an organization, including:

- Economic Downturns: Potential impacts of recessions or slow growth.
- Increased Competition: New market entrants or aggressive competitors.
- Technological Disruptions: Innovations that may render existing products or services obsolete.
- Regulatory Changes: Compliance risks and costs associated with new laws or policies.

Below are two frameworks suitable for analyzing the external environment.

1. *PEST Analysis*: PEST analysis is a strategic tool for examining macro-environmental factors that affect an organization's performance. By analyzing key external dimensions, businesses can better anticipate challenges, mitigate risks, and seize opportunities. These dimensions include the following:

   - Political: Government policies, tax regulations, trade restrictions, and political stability.
   - Economic: Growth trends, interest rates, inflation, exchange rates, and employment levels.
   - Sociocultural: Cultural norms, demographic shifts, lifestyle trends, and consumer behavior.
   - Technological: Innovation rates, emerging technologies, and digital transformation.

2. *Porter's Five Forces Analysis*: Porter's five forces framework assesses industry competition and profitability by examining five key forces:

   - Threat of New Entrants: The ease with which new competitors can enter the market, influenced by barriers such as high capital requirements, brand loyalty, and government regulations.
   - Bargaining Power of Suppliers: The extent to which suppliers can influence pricing and terms, potentially increasing costs for businesses.
   - Bargaining Power of Buyers: The ability of customers to demand lower prices or higher quality, affecting profitability.
   - Threat of Substitute Products or Services: The availability of alternative solutions that can reduce demand for an organization's offerings.

- Rivalry Among Existing Competitors: The intensity of competition within the industry, which can drive price wars, marketing battles, and reduced profit margins.

4.3.2.3 Cross-SWOT Analysis—A Deeper Dive

SWOT analysis helps organizations assess their internal and external environments, but Cross-SWOT analysis offers a deeper, more strategic perspective by examining the interplay between these elements.

Cross-SWOT analysis involves systematically pairing each SWOT element—Strengths, Weaknesses, Opportunities, and Threats—with the other three to uncover strategic insights. This matrix-based approach helps organizations develop well-rounded strategies by identifying how internal and external factors interact.

The steps to conduct a Cross-SWOT analysis are as follows:

1. Conduct a SWOT Analysis: Identify internal strengths and weaknesses, as well as external opportunities and threats.
2. Create a Cross-SWOT Matrix: Pair each SWOT element with the others, such as Strengths with Opportunities and Strengths with Weaknesses.
3. Analyze Strategic Implications: Assess each combination to determine potential actions. For example, leveraging strengths to capitalize on opportunities can inform a proactive growth strategy.

By using Cross-SWOT analysis, organizations gain a more comprehensive understanding of their strategic landscape and can craft innovative approaches to achieve their objectives. A thorough analysis considers all possible interactions between internal and external factors, including:

1. S-O (Strengths-Opportunities): Utilizing strengths to maximize opportunities.
2. S-T (Strengths-Threats): Leveraging strengths to counteract threats.
3. W-O (Weaknesses-Opportunities): Improving weaknesses to take advantage of opportunities.
4. W-T (Weaknesses-Threats): Minimizing weaknesses to reduce vulnerability to threats.
5. S-W (Strengths-Weaknesses): Identifying ways internal strengths can compensate for weaknesses.
6. O-T (Opportunities-Threats): Assessing how threats could impact opportunities and vice versa.

While all six combinations offer valuable insights, organizations typically focus on the first four as they directly connect internal capabilities with external conditions.

Figure 4.7 provides a Cross-SWOT analysis template.

Developing a Strategy: A Cross-SWOT analysis helps organizations formulate effective strategies by examining how strengths, weaknesses, opportunities, and threats interact. The following four key combinations provide strategic insights:

		External Factors	
		Opportunities	Threats
Internal Factors	Strengths	**Strengths × Opportunities** Maximize opportunities by making use of strengths.	**Strengths × Threats** Manage threats with your company's strengths.
	Weaknesses	**Weakness × Opportunities** Don't miss opportunities because of your company's weaknesses.	**Weaknesses × Threats** Avoid the worst-case scenario because of your company's weaknesses and threats.

Fig. 4.7 Cross-SWOT analysis template

1. Strengths × Opportunities
   - *Strategy*: Leverage strengths to capitalize on opportunities.
   - *Action*: Identify opportunities that align with core competencies and competitive advantages. Develop strategies to maximize these opportunities and strengthen market position.
2. Strengths × Threats
   - *Strategy*: Use strengths to mitigate threats.
   - *Action*: Assess how existing strengths can help counteract potential threats. Explore innovative approaches to turn threats into new opportunities.
3. Weaknesses × Opportunities
   - *Strategy*: Address weaknesses to seize opportunities.
   - *Action*: Identify internal limitations that hinder opportunity realization. Develop strategies to overcome these weaknesses and enhance organizational capabilities.
4. Weaknesses × Threats
   - *Strategy*: Minimize weaknesses to reduce vulnerability to threats.
   - *Action*: Focus on strengthening critical weaknesses that amplify the impact of threats. Consider defensive strategies such as cost reduction, restructuring, or divesting non-core activities.

4.3.2.4 Example: SWOT and Cross-SWOT Analysis for a Small Coffee Shop

SWOT Analysis: A small, independent coffee shop in a busy downtown area examines its strengths, weaknesses, opportunities, and threats:

A. *Strengths*

- High-Quality Coffee Beans: Sourced from reputable suppliers, offering unique flavors.
- Skilled Baristas: Experienced baristas capable of crafting various coffee beverages.
- Comfortable Atmosphere: Comfortable seating and friendly staff create a welcoming environment.
- Existence of Loyal Customers: A dedicated following that values quality and service.

B. *Weaknesses*

- Limited Seating Capacity: Small shop size restricts the number of customers.
- High Operating Costs: Rent, utilities, and premium ingredients result in significant expenses.
- Low Name Recognition: Less visibility compared to larger coffee chains.

C. *Opportunities*

- Growing Demand for Specialty Coffee: Increasing consumer interest in unique coffee experiences.
- Potential for Growth in Catering: Potential to expand revenue by offering catering for office meetings and events.
- Partnerships with Local Bakeries: Collaborating to provide a broader range of food items.
- Expansion of Online Orders and Delivery: Enhancing customer convenience through digital platforms.

D. *Threats*

- Economic Downturn: Potential declines in consumer spending during recessions.
- Increased Competition: The emergence of new coffee shops in the area.
- Rising Supply Costs: Fluctuating prices of coffee beans and ingredients.
- Changing Consumer Preferences: Evolving trends in coffee consumption.

Cross-SWOT Analysis: By combining elements of the SWOT analysis, the coffee shop can develop targeted strategies:

A. *S-O Strategies* (Strengths × Opportunities): Utilize strengths to maximize opportunities.

- High-quality coffee beans × Growing demand for specialty coffee → Develop a "premium experience menu" and raise the price range.
- Skilled baristas × Potential for growth in catering → Host corporate coffee workshops and tasting events to attract B2B clients and boost brand awareness.
- Comfortable atmosphere × Partnership with a local bakery → Position the café as a "locally produced, locally consumed" establishment that highlights regional identity.
- Presence of loyal customers × Expansion of online orders and delivery → Introduce a "subscription system (regular purchase)" for regular customers.

B. *S-T Strategies* (Strength × Threats): Avoid and mitigate threats by utilizing strengths.

- High-quality coffee beans × Increased competition → In order to differentiate from major companies, communicate "producer stories" and emphasize the individuality of the brand.
- Skilled baristas × Rising supply costs → Utilize barista skills to introduce "original blends" and "in-house roasting" to keep costs down while maintaining quality.
- Existence of loyal customers × Economic downturn → Introduce VIP membership and reward programs to increase customer spending and repeat rates.
- Comfortable atmosphere × Changing consumer preferences → Develop a dedicated "healthy choice" menu featuring caffeine-free, organic, and plant-based options.

C. *W-O Strategies* (Weakness × Opportunity): Overcome weaknesses while taking advantage of opportunities.

- Limited Seating Capacity × Expansion of online orders and delivery → Implement a streamlined mobile ordering system and introduce a "grab-and-go" concept to maximize efficiency.
- High operating costs × Potential for growth in catering → Develop a "low-cost, high-margin coffee service for offices" to boost revenue while minimizing fixed costs.
- Low name recognition × Partnering with a local bakery → Plan joint events and limited collaboration products, and leverage each other's customer base to increase brand recognition.
- Low name recognition × Expansion of online orders and delivery → Utilize SNS and word of mouth to conduct online advertising campaigns.

D. *W-T Strategies* (Weakness × Threats): Mitigate the risk of overlapping weaknesses and threats.

- Limited seating capacity × Increased competition → Introduce a "coffee stand" model, reducing reliance on dine-in customers and increasing turnover.

- High operating costs × Economic downturn → Negotiate better supplier contracts and implement cost-saving measures (e.g., energy-efficient equipment, waste reduction).
- Low name recognition × Changing consumer preferences → Emphasize health consciousness and sustainability to develop new segments.
- Low name recognition × Increased competition → Launch a customer referral program offering discounts or free drinks for bringing new customers.

Summary: This Cross-SWOT Analysis clarified the following strategies:

1. Differentiation Strategy (S-O, S-T)
    - Premium customer experience (premium menu, producer stories, skilled baristas)
    - Unique blends and in-house roasting to maintain quality while managing costs
    - Subscription model to strengthen customer retention
    - VIP membership and reward programs to drive repeat purchases
    - Healthy choice menu to cater to evolving preferences
2. Growth Strategy (W-O, some S-O)
    - Digitalization (mobile ordering, SNS marketing, word-of-mouth campaigns)
    - Collaboration projects (local bakery partnerships, joint events)
    - Catering expansion (corporate coffee workshops, B2B coffee services)
    - Grab-and-go concept to maximize efficiency with limited seating
3. Risk Mitigation Strategy (W-T, some S-T)
    - Cost reduction (supplier negotiations, waste reduction, energy-efficient equipment)
    - Capturing new demand (health-conscious and sustainability-focused branding)
    - Optimizing store model (coffee stand concept to reduce dine-in dependency)
    - Customer referral programs to combat low name recognition

4.4 Innovation and Creativity

A strong understanding of innovation theory can significantly enhance decision-making and problem-solving. While the concept of innovation has evolved over the past century, its core principles remain essential. Exploring renowned theories and real-world examples can inspire creative thinking and encourage fresh approaches to challenges.

This section examines key innovation theories and their relevance to decision-making and problem-solving. It concludes with an in-depth exploration of design thinking, a widely used methodology that provides practical tools for addressing complex problems.

4.4.1 *Definition and Theories of Innovation*

Studying innovation theories and real-world examples can significantly enhance decision-making and problem-solving in the following ways:

1. Stimulating Creative Thinking: Innovation theories encourage divergent thinking and inspire unconventional problem-solving approaches. Analyzing case studies of successful innovations can spark new ideas and solutions.
2. Fostering a Growth Mindset: Learning from innovators highlights the value of perseverance and resilience. Many groundbreaking ideas fail before succeeding, making failure a crucial step in the innovation process. Embracing this mindset fosters continuous learning and experimentation.
3. Cultivating Systems Thinking: Innovation theories emphasize the interconnectedness of systems. Understanding these connections helps identify root causes of issues and design comprehensive, sustainable solutions.
4. Enhancing Problem-Solving Skills: Methodologies like design thinking provide a structured, iterative framework for problem-solving. This approach promotes experimentation, refinement, and continuous improvement.
5. Strengthening Decision-Making: Innovation theories stress the importance of data-driven decisions and proactive trend analysis. They enable organizations to make informed choices that align with long-term goals and provide a competitive advantage.

4.4.1.1 What is Innovation?

Innovation is the successful introduction of new ideas, methods, or products. It involves taking calculated risks and experimenting with novel approaches to solve problems or create opportunities. Innovation can range from incremental improvements to disruptive technologies that redefine industries.

Key aspects of innovation include:

- *Idea Generation*: New ideas emerge through brainstorming, research, and creative techniques.
- *Implementation*: Promising ideas require strategic planning, resources, and effective execution to become reality.
- *Value Creation*: True innovation generates value by increasing revenue, improving efficiency, or delivering social benefits.
- *Risk-Taking*: Innovation involves calculated risks. While not every idea succeeds, learning from failures is essential for growth.

Innovation is crucial for businesses to stay competitive and for society to progress. It drives the development of new products, services, and processes that enhance quality of life. Organizations that cultivate a culture of innovation can improve customer satisfaction, drive growth, and build a sustainable future.

4.4.1.2 A Modern Perspective on Innovation

Joseph A. Schumpeter, a renowned economist, laid the foundation for modern innovation theory (Schumpeter, 1912, 1942). He defined *innovation* as "*the introduction of new combinations*" of production factors, such as new products, new production methods, new markets, new sources of supply, and new organizational structures.

While Schumpeter's definition remains influential, the concept of innovation has evolved to encompass broader dimensions:

- *Incremental vs. Radical Innovation*: Modern innovation encompasses both incremental improvements to existing products or processes and radical breakthroughs that disrupt industries, such as the advent of smartphones or renewable energy solutions.
- *Innovation Ecosystems*: Collaboration between businesses, academia, government, and society has become a driving force behind innovation, creating ecosystems where ideas flourish.
- *Social and Environmental Impact*: Today, innovation extends beyond economic growth, addressing social, environmental, and ethical challenges such as sustainability and equity.

Despite these evolving perspectives, Schumpeter's core idea that innovation is a key driver of economic growth and societal progress remains highly relevant. His work continues to inspire contemporary thinking on innovation and entrepreneurship.

Building on Schumpeter's foundation, Peter Drucker, a renowned management consultant and author, expanded the understanding of innovation with a more practical, application-oriented approach (Drucker, 1954, 1973, 1985). Drucker viewed innovation as the deliberate effort to create meaningful change, enhancing an enterprise's economic or social potential. He saw innovation as a discipline that could be learned and practiced, not just a matter of luck or genius.

Key insights from Drucker's work include:

- Innovation as a Systematic Process: Drucker emphasized that innovation is not a random occurrence but a structured process. He outlined steps for identifying opportunities, developing ideas, and effectively implementing them.
- Customer-Centric Innovation: Drucker argued that innovation must be market-driven, rooted in a deep understanding of customer needs, and focused on creating value. For example, Apple has excelled by prioritizing user experience, demonstrating how customer-centric innovation drives long-term success.
- Innovation as a Competitive Advantage: Drucker viewed innovation as essential for maintaining a competitive edge. Companies that continuously introduce new products, services, or processes differentiate themselves and thrive in dynamic markets.
- Innovation as a Social Responsibility: Drucker recognized innovation's potential to contribute to societal well-being by driving economic growth, creating jobs, and improving quality of life.

Drucker's insights remain a cornerstone of management and innovation theory. His work offers practical guidance for organizations striving to foster innovation, meet customer needs, and achieve sustainable growth.

4.4.1.3 Key Theories of Innovation

Understanding innovation theories is crucial for organizations seeking to develop effective strategies and drive sustainable growth. Several key innovation theories shape modern business strategies:

1. *Disruptive Innovation*: Clayton Christensen's theory of disruptive innovation describes how emerging technologies or business models start in niche markets and eventually disrupt established industries. Examples include Uber's impact on transportation, Airbnb's transformation of hospitality, and Spotify's disruption of the music industry.
2. *Sustaining Innovation*: Sustaining innovation improves existing products or services to meet the evolving needs of current customers. Examples include the continuous upgrades of the iPhone or the introduction of new car models with enhanced features.
3. *Radical Innovation*: Radical innovation involves breakthrough technological advancements that create entirely new markets or industries. Historical examples include the invention of the automobile, the personal computer, and the Internet.
4. *Incremental Innovation*: Incremental innovation focuses on minor, continuous improvements to existing products or processes. Examples include regular software updates, enhanced product features, and optimized workflows.

Beyond these theories, additional frameworks offer structured approaches to innovation:

- *Architectural Innovation*: Architectural innovation involves recombining existing technologies or components in new ways to create novel products or services. For instance, the smartphone integrates a phone, camera, music player, and Internet access into one device.
- *Modular Innovation*: Modular innovation creates standardized components that can be combined and reconfigured into new products or services. Examples include Lego bricks and modular furniture, which offer flexibility, customization, and easy upgrades.
- *Process Innovation*: Process innovation enhances the efficiency and effectiveness of business processes by streamlining workflows, automating tasks, and improving quality control. Lean manufacturing, Six Sigma, and automation are prime examples.

To cultivate innovation, organizations can leverage the following frameworks:

- *TRIZ* (*Theory of Inventive Problem-Solving*): A systematic, structured approach to solving complex problems and fostering innovative thinking.
- *Design Thinking*: A human-centered methodology focusing on empathy, ideation, prototyping, and testing to develop creative solutions.
- *Blue Ocean Strategy*: A strategy that seeks to create uncontested market space, moving away from direct competition by focusing on differentiation and value innovation.
- *Open Innovation*: A collaborative approach that involves leveraging external partnerships, such as startups, academia, or suppliers, to drive innovation.

The following will delve into Disruptive Innovation and Blue Ocean Strategy, exploring their principles, real-world applications, and strategic impact. These influential theories provide actionable insights for businesses seeking to stay ahead in competitive markets.

Christensen's Innovation Theory—Disruptive Innovation: Clayton Christensen, a renowned innovation theorist, revolutionized our understanding of innovation by introducing the concept of *disruptive innovation* in his seminal work, *The Innovator's Dilemma* (Christensen, 1997).

Traditionally, innovation has been categorized based on technological advancement:

- *Incremental Innovation*: Gradual, continuous improvements to existing products or processes.
- *Radical Innovation*: Significant breakthroughs that introduce new technologies or entirely novel business models.

Christensen proposed a new classification that focuses on the market impact of innovation:

- *Sustaining Innovation*: Enhancements to existing products or services aimed at satisfying the needs of current customers.
- *Disruptive Innovation*: Innovations that initially cater to niche markets with underserved needs but eventually disrupt and reshape established industries.

Key characteristics of disruptive innovation include:

- Niche Market Focus: Disruptive innovations often begin in niche markets, targeting customers who are overlooked or underserved by existing products or services.
- Simple and Affordable: These innovations are typically more affordable, accessible, and easier to use than high-end alternatives, making them attractive to new market segments.
- Performance Overkill: Disruptive innovations may initially fall short of the performance standards demanded by mainstream markets. However, they improve and evolve over time to challenge and surpass established products.
- Resistance from Established Firms: Established companies often resist disruptive innovations because they can threaten existing revenue streams and business models.

Disruptive innovations succeed for several reasons, often capitalizing on the limitations of established organizations:

- Organizational Inertia: Large, established companies often focus on optimizing their core operations, making them slow to adapt to emerging trends.
- Resource Allocation Bias: Businesses prioritize investments in sustaining innovations that enhance existing products, inadvertently neglecting opportunities for disruption.
- Market Perceptions: Disruptive innovations are often dismissed as inferior or niche solutions during their early stages, allowing them to grow unnoticed until they pose a significant threat.

Understanding the dynamics of disruptive innovation equips organizations to:

- Identify emerging threats and opportunities early.
- Embrace a culture of experimentation and risk-taking.
- Balance investments between sustaining and disruptive innovations to remain competitive.

Examples of disruptive innovation include:

- Netflix disrupted the DVD rental industry with its subscription-based streaming model.
- Tesla introduced electric vehicles that initially targeted high-end markets but later expanded to broader segments.
- Airbnb revolutionized the hospitality industry by providing affordable, peer-to-peer lodging options.

Organizations can leverage disruptive innovations to drive long-term growth and thrive in an ever-changing marketplace by fostering innovation and remaining agile.

Blue Ocean Strategy—A Path to Uncontested Market Space: In today's highly competitive business landscape, most companies focus on outperforming rivals within existing markets—what Kim and Mauborgne (2005) describe as "red oceans." However, Blue Ocean Strategy offers an alternative path: creating entirely new markets, or "blue oceans," where competition becomes irrelevant. This strategic framework encourages companies to innovate and redefine market boundaries, unlocking new demand and creating sustainable growth opportunities.

Value Innovation: At the heart of Blue Ocean Strategy lies the concept of value innovation—simultaneously pursuing differentiation and low cost to deliver exceptional value to customers while reducing costs.

The framework introduces the Four Actions Framework, guiding organizations to reshape their industry's value curve:

- *Eliminate*: Identify and eliminate factors that are no longer relevant or add little value to customers.
- *Reduce*: Lower investment in factors that exceed customer needs.
- *Raise*: Enhance factors that significantly impact customer satisfaction.
- *Create*: Introduce entirely new factors or features that redefine the industry.

Key benefits of Blue Ocean Strategy are:

1. Breaking Free from Competition: Instead of battling over market share, companies explore untapped opportunities, creating uncontested spaces for growth.
2. Challenging Industry Norms: By questioning conventional wisdom, businesses uncover new possibilities that competitors may overlook.
3. Creating New Demand: Developing innovative products or services attracts new customer segments, expanding the market.
4. Achieving Long-Term Growth: Organizations that successfully implement Blue Ocean Strategy often experience sustained profitability by leading new market spaces.

Several organizations have successfully leveraged Blue Ocean Strategy to reshape their industries:

- Cirque du Soleil: Transformed traditional circus entertainment by eliminating costly animal acts and creating a sophisticated, theatrical experience targeting adults.
- Nintendo Wii: Redefined the gaming market by focusing on motion-based controls and family-friendly games, appealing to a broader audience than traditional gamers.
- Yellow Tail Wine: Simplified wine choices for non-expert consumers by reducing complexity and creating an approachable, affordable brand.

To implement Blue Ocean Strategy effectively, organizations should:

1. Analyze the Industry: Identify pain points, unmet customer needs, and inefficiencies in the current market.
2. Reconstruct Market Boundaries: Use the Four Actions Framework to redefine customer value.
3. Test and Validate: Experiment with innovative ideas to ensure that they resonate with target audiences.
4. Build Organizational Support: Foster a culture of innovation and collaboration to sustain blue ocean initiatives.

By embracing Blue Ocean Strategy, companies can move beyond cutthroat competition and create unique value propositions that drive growth and innovation. When combined with the principles of value innovation, this framework provides a powerful tool for achieving long-term success and gaining a competitive advantage in an ever-changing marketplace.

4.4.1.4 Analyzing Uber: A Case Study in Disruptive Innovation

Uber, a global technology company, has transformed the transportation industry by leveraging technology and creating new market opportunities. By offering app-based ride-hailing services, Uber has challenged traditional taxi services and reshaped customer expectations for convenience and efficiency.

Uber as a Disruptive Innovation: As theorized by Clayton Christensen, disruptive innovation involves introducing new products or services that initially target niche markets but eventually disrupt established industries. Uber exemplifies this concept in the following ways:

1. *Niche Market Focus*: Uber initially targeted specific market segments, such as high-end users with UberBLACK (a premium service with luxury vehicles and professional drivers) and urban professionals looking for a reliable alternative to taxis.
2. *Performance Overkill*: In its early stages, Uber's service lacked the infrastructure and reliability of traditional taxi companies. However, its convenience, app-based interface, and transparent pricing appealed to customers dissatisfied with existing options.
3. *Overcoming Resistance*: Traditional taxi companies resisted Uber's entry into the market, often lobbying for regulatory actions. Despite this, Uber's focus on customer convenience and competitive pricing led to rapid adoption.
4. *Technology-Driven Transformation*: Features like real-time ride tracking, cashless payments, and dynamic pricing set Uber apart and created a new value proposition for riders and drivers.

Uber's disruptive innovation lies in its ability to leverage technology to meet unaddressed customer needs while gradually improving its service quality and expanding its market reach.

Does Uber Follow a Blue Ocean Strategy? While Uber demonstrates elements of a Blue Ocean Strategy, its overall approach aligns more closely with disruptive innovation.

1. *Blue Ocean Elements in Uber's Strategy*:
   - Creating New Demand: Uber challenged the traditional taxi industry by introducing a convenient, app-based ride-hailing model that appealed to tech-savvy and underserved customers.
   - Value Innovation: Uber redefined customer expectations by eliminating the need for physical taxi stands, reducing wait times, and offering flexible pricing.
2. *Limitations of Uber's Blue Ocean Strategy Approach*:
   - Unlike traditional Blue Ocean examples, Uber's market entry intensified competition within the transportation industry. Rival companies like Lyft, Grab, and Bolt have emerged, creating a highly competitive landscape rather than an uncontested market space.
   - While Uber shares principles with Blue Ocean Strategy—such as challenging industry conventions and innovating its value curve—it primarily operates within a competitive environment driven by disruptive innovation.

Uber's rise highlights the importance of combining innovative strategies with effective execution:

- Embracing Technology: Uber's success has been driven by its app-based platforms, real-time tracking, and AI-driven algorithms.
- Customer-Centric Approach: Uber solves pain points such as unreliable service, opaque pricing, and payment inconvenience.
- Adaptive Strategy: To capture additional market segments, the company diversified its offerings, such as UberPOOL for shared rides and UberEATS for food delivery.

Uber's success is a powerful example of how disruptive innovation can transform an industry. While it incorporates elements of a Blue Ocean Strategy, its approach addresses customer dissatisfaction with traditional services and leverages technology to create new market opportunities. Organizations can learn from Uber's ability to identify unmet needs, embrace innovation, and execute strategies that balance differentiation and cost efficiency. By combining these elements, companies can navigate competitive markets and position themselves for sustained growth.

4.4.2 Design Thinking

Design thinking is an iterative, human-centered problem-solving methodology that originated in the fields of architecture and urban planning. Initially focused on physical design, design thinking has since been widely adopted across various creative disciplines due to its ability to generate innovative and practical solutions through active problem discovery and empathy.

Design thinking gained prominence in the early 2000s after several significant milestones:

- The founding of the Hasso Plattner Institute of Design (d.school) at Stanford University in 2005 brought academic recognition to the approach.
- A special "Design Thinking" issue published by BusinessWeek the same year highlighted its potential for business applications.
- In 2008, Tim Brown, CEO of IDEO, further popularized the concept by introducing "IDEO Design Thinking" in Harvard Business Review (Brown, 2008).

Unlike the scientific method, which focuses on objective data, design thinking emphasizes empathy—a deep understanding of users' needs, desires, and frustrations. It achieves this through techniques such as:

- User interviews and surveys to gather qualitative insights.
- Ethnographic research to observe user behavior in real-world contexts.

By prioritizing empathy, design thinking ensures that solutions are both practical and emotionally resonant with users.

Design thinking is an iterative process that encourages revisiting and refining ideas throughout development. Its flexibility allows teams to:

- Challenge Assumptions: Question initial beliefs or preconceived notions about the problem.
- Redefine Problems: Incorporate new insights from user research to reframe challenges effectively.
- Co-evolve Solutions: Continuously improve designs based on iterative user feedback.

This adaptability fosters creative, user-focused solutions that dynamically evolve as teams gain a deeper understanding of the problem and user needs.

The design thinking process consists of five stages:

1. *Empathize*: Conduct user research to identify needs, motivations, and pain points.
2. *Define*: Synthesize insights from research to articulate a clear problem statement.
3. *Ideate*: Brainstorm a wide range of creative solutions.
4. *Prototype*: Build simple, low-fidelity models or mockups of potential solutions.
5. *Test*: Gather user feedback on prototypes and refine the design accordingly.

While this five-stage framework is widely associated with Stanford's d.school, design thinking principles have been applied across disciplines for decades.

Design thinking is particularly effective in addressing complex, ambiguous problems where understanding user needs and emotions is critical. Its emphasis on empathy, creativity, and iteration makes it a valuable tool for solving challenges that require human-centered solutions.

However, it is essential to recognize that the scientific method is more suitable for situations requiring rigorous testing, objective data analysis, and hypothesis validation. By understanding the strengths of both approaches, you can select the most effective framework for a given problem.

4.4.2.1 Key Frameworks for Design Thinking

Design thinking is a human-centered approach to innovation that utilizes various frameworks and tools to guide the creative process. Below are some popular frameworks and techniques for each stage:

1. Empathize Stage

    - User Interviews: Conduct in-depth interviews with potential users to understand their needs, pain points, and motivations.
    - Observation: Observe users in their natural environment to gain insights into their behaviors and interaction patterns.
    - Empathy Maps: Create visual representations of users' thoughts, feelings, actions, and pain points.
    - Customer Journey Maps: Map out users' journeys to achieve their goals, identifying key touchpoints and areas for improvement.
    - User Personas: Develop fictional representations of target users to better understand their needs and motivations.

2. Define Stage
   - Problem Statement: Clearly articulate the core problem or challenge to be addressed.
   - How Might We (HMW) Questions: Formulate open-ended questions to spark creative thinking and generate potential solutions.
   - Affinity Mapping: Group-related ideas to identify patterns and themes.
   - Customer Journey Maps: Revisit this tool to pinpoint pain points and opportunities for improvement in the user journey.
3. Ideate Stage
   - Brainstorming: Encourage participants to explore ideas freely without limitations.
   - Mind Mapping: Visually organize ideas to explore connections and relationships.
   - SCAMPER: Apply this checklist (Substitute, Combine, Adapt, Modify, Put to another use, Eliminate, Reverse) to generate innovative ideas.
4. Prototype Stage
   - Paper Prototyping: Create quick, low-fidelity prototypes to test ideas.
   - Digital Prototyping: Use tools to build interactive prototypes.
   - Physical Prototyping: Construct tangible prototypes to validate functionality and user experience.
   - Role-playing: Simulate user interactions to identify potential usability issues.
5. Test Stage
   - User Testing: Observe users interacting with prototypes and collect feedback.
   - A/B Testing: Compare design variations to determine the most effective solution.
   - Surveys: Gather quantitative data on user preferences and satisfaction.
   - Iterative Design: Continuously refine the design based on feedback and testing outcomes.

These frameworks are adaptable tools that can be customized to suit the specific needs of a project and team. The essence of design thinking lies in creativity, experimentation, user empathy, and collaboration.

4.4.2.2 Using Frameworks for Decision-Making: A Design Thinking Approach

Design thinking follows a two-phase process: divergent thinking to generate many ideas, followed by convergent thinking to narrow options and identify the most promising solutions. This iterative process enables the exploration of various possibilities before refining and implementing the best ideas. Below are key frameworks used in each phase:

- *Divergent Thinking Frameworks*: Brainstorming, Mind Mapping
- *Convergent Thinking Frameworks*: Affinity Diagrams, Two-Axis Charts

The following provides a brief overview of these frameworks, some of which were also discussed in Sect. 4.3.1.

A. *Brainstorming*: Brainstorming is a classic technique for quickly generating a wide range of ideas. To maximize its effectiveness, follow these principles:

   - Quantity over Quality: Prioritize producing many ideas initially without worrying about feasibility.
   - Defer Judgment: Avoid criticism or evaluation during the idea generation phase to encourage creativity.
   - Embrace Wild Ideas: Welcome unconventional or bold concepts to challenge assumptions and spark innovation.
   - Build on Ideas: Expand or combine existing ideas to create innovative solutions.

B. *Mind Mapping*: Mind mapping is a visual tool that hierarchically organizes ideas and illustrates their relationships. It begins with a central theme or concept, branching into related ideas. Unlike brainstorming, which can sometimes lack structure, mind mapping supports systematic exploration by enabling both:

   - Vertical Thinking: A deeper dive into individual ideas.
   - Horizontal Thinking: Discovery of new connections between ideas.

   Mind Mapping combines these two types of thinking to enable a more thorough and structured exploration of possibilities.

C. *Affinity Diagrams*: Affinity diagrams are powerful tools for organizing and categorizing large sets of information. They are particularly useful for:

   - Identifying patterns and trends: Grouping similar ideas helps reveal underlying themes.
   - Prioritizing ideas: Categorizing ideas helps focus attention on the most important or relevant ones.
   - Generating new insights: Structuring information can lead to unexpected connections and innovative solutions.

D. *Two-Axis Charts*: Two-axis charts, such as scatter plots, line graphs, and bubble charts, allow for the comparison of two variables on different scales or units. They are helpful for:

   - Identifying correlations and trends: Visualizing relationships between two data sets.
   - Making data-driven decisions: Using visualized patterns to make informed choices.

By effectively leveraging these frameworks, you can enhance decision-making, foster creativity, and achieve better outcomes. These versatile tools can be applied to various challenges, from personal decisions to complex business strategies.

As an example, let's apply divergent and convergent thinking to design a summer menu for a sushi restaurant.

Summer Menu Planning with Brainstorming and an Affinity Diagram: Imagine a conveyor belt sushi restaurant conducting a brainstorming session to develop a vibrant summer menu. During the session, the team generated the following ideas:

- Seasonal Ingredients: Highlight summer's bounty with seasonal ingredients like seared bonito, horse mackerel, conger eel, tomatoes, cucumbers, and eggplant.
- Summer Vegetables: Showcase seasonal vegetables in refreshing dishes.
- Chilled Dishes: Introduce cold options such as chilled wheat noodles, ramen, and soups.
- Innovative Creations: Experiment with unique dishes like pineapple or watermelon sushi.
- Seafood Bowls: Provide popular seafood bowls served with warm vinegared rice.
- Hand Rolls: Introduce a DIY hand-roll station where customers can assemble their own sushi.
- Enhanced Side Dishes: Expand the side dish selection with options like shrimp, vegetable tempura, and egg dishes.
- Sweet Treats: Offer summery desserts such as shaved ice, ice cream, and fruit punch.
- Health-Conscious Options: Cater to health-conscious diners with low-calorie and nutritious choices.
- Kid-Friendly Delights: Design cute, tasty dishes to appeal to young customers.
- Social Media-Worthy: Create visually stunning dishes that encourage customers to share on social media.
- Local Flavor: Incorporate regional ingredients and specialties to add a unique, local touch.

Using an Affinity Diagram: To structure their ideas, the team categorized and prioritized them using an affinity diagram. They categorized the brainstormed ideas into four main groups: "Signature Dishes," "Light & Healthy Options," "Kid-Friendly Choices," and "Seasonal Ingredients."

Within these groups, ideas were prioritized based on their profitability potential, as shown in Fig. 4.8. This structured approach allowed the team to:

- Group Similar Ideas: Identify relationships among the concepts.
- Focus on Profitability: Prioritize dishes that align with customer preferences while maximizing revenue potential.
- Strategically Combine Ideas: Mix high-potential ideas to create an appealing, cohesive menu.

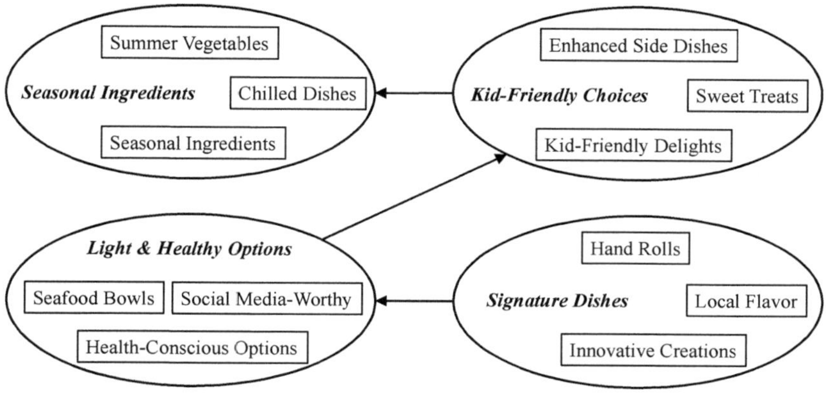

Fig. 4.8 An affinity diagram prioritizing ideas from a profitability standpoint

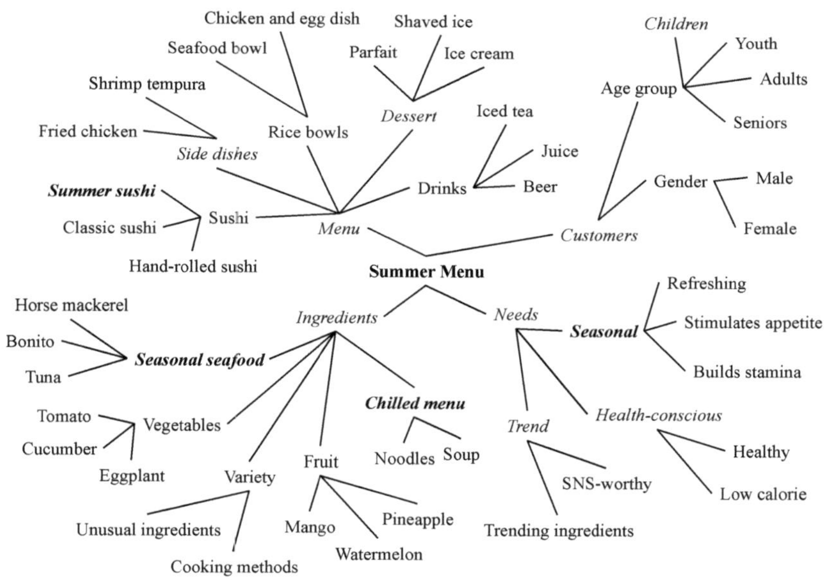

Fig. 4.9 A mind map for planning a summer menu at a conveyor belt sushi restaurant

By combining brainstorming with affinity diagrams, the restaurant can craft a summer menu that highlights seasonal flavors, appeals to diverse customers, and maximizes profitability.

Summer Menu Planning with Mind Mapping and a Two-Axis Chart: A mind map, such as the one in Fig. 4.9, helps the restaurant visually map out and structure ideas for a vibrant summer menu. It highlights a range of possibilities and provides a structured overview of potential dishes and concepts.

Key Insights from the Mind Map: Analyzing the mind map reveals several key considerations for planning the summer menu:

- Family-Friendly Options: Kid-friendly dishes are in high demand as more families dine out during summer vacations.
- Social Media Appeal: Visually appealing dishes attract customers and drive social media engagement.
- Diverse Menu: Offering a variety of side dishes and desserts enhances customer satisfaction.
- Light and Refreshing Choices: Lighter, refreshing dishes are essential for hot weather.
- Energizing Options: Some customers seek dishes that boost energy and stamina.
- Seasonal Delights: Incorporating ingredients like eel adds a unique, seasonal touch to the menu.

From these insights, the restaurant can develop specific menu ideas:

- Youth Appeal: Bright, colorful seafood bowls and innovative sushi combinations.
- Seasonal Specialties: Highlighting dishes such as eel-based offerings and fatty tuna nigiri.
- Summer Desserts: Watermelon-based treats and tropical parfaits to cater to summer cravings.

Optimizing the Menu with a Two-Axis Chart: The restaurant can use a two-axis chart to further refine the summer menu. Menu items from the mind map are mapped onto a two-axis chart for further analysis, as shown in Fig. 4.10.

- Horizontal Axis: Represents customer age groups, ranging from children to adults.
- Vertical Axis: Compares seasonally inspired dishes with standard menu items.

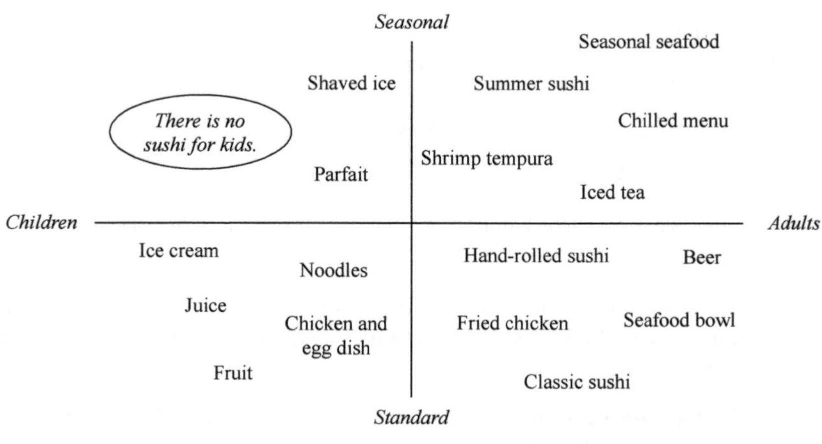

Fig. 4.10 A two-axis chart highlighting market gaps and potential opportunities for the summer menu

This visual tool helps uncover menu gaps and new opportunities. For example, the chart may highlight the need for a family-set menu featuring sushi, side dishes, and drinks tailored to children's preferences.

4.4.2.3 Design Thinking in Action: Successful Case Studies

Numerous companies have leveraged design thinking to transform industries and enhance user experiences. Here are some notable examples:

- Apple: Apple serves as a leading example of design thinking in action. The company consistently reshapes industries by prioritizing user experience and intuitive design. Iconic products such as the iPhone and iPad exemplify this approach. By deeply understanding user needs and iteratively refining its products, Apple has redefined consumer electronics.
- IBM: As a technology giant, IBM has embraced design thinking as a core business strategy. By investing in design teams and implementing design thinking methodologies, IBM has tackled complex challenges and delivered innovative solutions for its clients. This approach has resulted in increased customer satisfaction and improved business outcomes.
- GE Healthcare: Despite having an advanced medical imaging device, GE Healthcare applied design thinking to uncover user pain points. By prioritizing the needs of patients and healthcare providers, GE Healthcare developed innovative solutions that enhanced patient care and streamlined healthcare delivery.
- Netflix: Netflix transformed the entertainment industry by leveraging design thinking to create a seamless and personalized streaming experience. Netflix has ensured its enduring popularity and relevance by continuously innovating and adapting to user preferences, guided by design thinking principles.
- Airbnb: Facing significant early challenges, Airbnb successfully applied design thinking principles to pivot its business model. By enhancing the user experience and fostering a global community of hosts and travelers, Airbnb established itself as a global leader in the travel and hospitality industry.

4.4.2.4 The Benefits and Challenges of Design Thinking

Benefits of Design Thinking: Design thinking is a systematic, human-centered framework for innovation and problem-solving. This methodology helps organizations identify and seize new opportunities, foster stronger customer relationships, and boost overall performance. Key benefits include:

- Customer-Centric Approach: Focuses on the needs and experiences of users, creating solutions that truly resonate with customers.
- Innovation and Creativity: Encourages divergent thinking to explore unconventional and novel ideas for addressing challenges.

- Reduced Risk and Accelerated Time-to-Market: Rapid prototyping and iterative testing enable organizations to validate ideas early, minimize risks, and speed up development cycles.
- Improved Collaboration and Teamwork: Fosters interdisciplinary collaboration, promoting diverse perspectives and shared ownership in problem-solving.
- Competitive Advantage: Helps organizations differentiate their offerings by delivering user-focused, innovative solutions, keeping them ahead in competitive markets.

Challenges of Design Thinking: While design thinking offers significant advantages, it poses some difficulties. Common limitations include:

- Overemphasis on User-Centricity: An exclusive focus on user needs might overlook broader business objectives or societal impacts.
- Lack of Critical Thinking and Systemic Analysis: Neglecting root causes and systemic complexities can lead to superficial solutions.
- Oversimplification of Complex Problems: Multifaceted issues may require deeper analysis beyond the scope of design thinking.
- Lack of Rigor and Evidence-Based Approaches: Heavy reliance on creativity and user feedback may result in decisions that lack empirical support
- Overreliance on Creativity and Innovation: Excessive focus on generating ideas can overshadow practical considerations such as feasibility, scalability, or long-term sustainability.

Balancing Design Thinking with Other Approaches: Addressing these challenges requires integrating design thinking with complementary methodologies to create a more balanced, comprehensive approach. In particular, blending creativity with analytical rigor can help organizations achieve a more holistic, impactful approach to innovation and problem-solving. These methodologies include:

- Systems Thinking: Considers interdependencies and broader systemic factors.
- Critical Design: Questions assumptions and explores alternative perspectives.
- Evidence-Based Approaches: Integrates data-driven insights to inform decision-making.

Beyond Design Thinking: While design thinking is a powerful tool, it alone cannot guarantee success. Several other factors play a crucial role in driving innovation:

- Strong Leadership: Visionary leaders inspire and guide teams toward innovative goals.
- Innovative Culture: Fosters creativity, risk-taking, and experimentation.
- Effective Execution: Ensures that ideas are implemented efficiently and effectively.
- Financial Resources: Secures adequate funding for research, development, and marketing efforts.
- Strong Brand Reputation: Builds trust and attracts customers and investors.
- Adaptability: Enables organizations to respond to evolving market conditions, consumer preferences, and emerging technologies.

By combining design thinking principles with these critical factors, organizations can cultivate a culture of innovation, foster long-term growth, and thrive in today's dynamic business landscape.

4.5 End-of-Chapter Problems

4.5.1 Problem 4.1: Reasoning Methods

We face countless decisions daily, even seemingly trivial ones like choosing what to eat for lunch. However, making optimal choices requires thoughtful consideration, even in these simple scenarios.

Task: Imagine entering a restaurant with a diverse lunch menu. How can you use *inductive*, *deductive*, and *abductive reasoning* to make an informed and satisfying lunch selection?

4.5.2 Problem 4.2: Knowledge Management

Tacit knowledge refers to unspoken, experiential insights deeply embedded in personal experience, intuition, and cultural context. Unlike explicit knowledge, it is difficult to articulate, codify, and share. Examples of tacit knowledge within organizations include:

- Craftsmanship: Nuanced skills and techniques honed over years, such as those of master chefs or artisans.
- Customer Intuition: Anticipating customer needs and preferences through subtle cues and unspoken emotions.
- Problem-Solving Expertise: Resolving complex issues based on intuition and past experiences.
- Organizational Culture: Shared values, norms, and behaviors conveyed through unspoken traditions.
- Personal Relationships: Trust and rapport built through interpersonal connections, influencing collaboration and decision-making.
- Leadership Abilities: Inspiring and motivating teams through actions, charisma, and intuition rather than formal instruction.
- Innovative Thinking: Generating novel ideas and solutions through intuitive leaps and unconscious connections.
- Product Intuition: Anticipating customer needs and designing innovative products by leveraging deep insights into behavior and market trends.

Task: Effectively harnessing tacit knowledge is essential for fostering organizational knowledge creation and management. However, converting it into explicit knowledge and facilitating its sharing presents significant challenges:

1. What are the key challenges in utilizing tacit knowledge within an organization?
2. How can organizations overcome these challenges to leverage their tacit knowledge assets effectively?

4.5.3 Problem 4.3: Logic Tree

Choosing a career is one of life's most significant decisions. A logic tree is a powerful tool for structuring and analyzing complex choices, making it especially useful for career planning. Below is a step-by-step guide for students to create a logic tree for their career path:

1. Identify the Main Decision: Define your primary decision as the trunk of the tree: "What career path should I select?"
2. Create the First Branches: Divide the main decision into key categories, such as industry, job role, work environment, and lifestyle.
3. Add Secondary Branches: Explore specific options within each category (e.g., industries like technology, healthcare, or finance).
4. Consider Tertiary Branches: Delve deeper into particular career paths (e.g., software engineer, data analyst, or cybersecurity specialist).
5. Assess Each Branch: Weigh the pros and cons of each tertiary branch, taking into account your interests, skills, and long-term goals.
6. Choose a Path: Select the career path that best aligns with your aspirations and vision for the future.

Here are some tips for creating effective logic trees:

- Keep It Clear: Do not overcomplicate the tree with too many branches. Focus only on the most relevant categories and options.
- Be Realistic: Consider your skills, limitations, and available opportunities.
- Stay Flexible: Recognize that your career path may evolve, so remain open to change.
- Visualize Your Future: Picture yourself in various roles to help clarify which paths resonate most with you.

Task

1. Create Your Logic Tree: Develop a logic tree for your job search using the steps above. Begin with "Career Path" as the root node. For the first branches, consider categories such as Industry, Role, Work Environment, and Lifestyle.
2. Imagine Your Career Path: Use the logic tree to envision your career journey. A career path represents the progression of organizational roles and outlines the steps needed for advancement.

4.5.4 Problem 4.4: SWOT and Cross-SWOT Analysis

Conduct a SWOT and Cross-SWOT analysis to develop a summer menu plan for a conveyor belt sushi restaurant. The following is a SWOT analysis of the restaurant:

1. Strengths:
   - Brand Recognition: Strong brand loyalty and reputation.
   - Efficient Operations: Well-established and streamlined processes.
   - Fresh Ingredients: High-quality, fresh ingredients.
2. Weaknesses:
   - Limited Menu Variety: Repetitive menu offerings.
   - Seasonal Fluctuations: Decreased foot traffic during certain times of the year.
   - Price Sensitivity: Customers may prioritize affordability over premium options.
3. Opportunities:
   - Food Trends: Incorporate popular trends like fusion cuisine and plant-based options.
   - Seasonal Ingredients: Leverage fresh, seasonal ingredients.
   - Online Ordering and Delivery: Expand service channels for added convenience.
4. Threats:
   - Economic Downturn: Reduced consumer spending.
   - Increased Competition: New restaurant openings.
   - Supply Chain Disruptions: Potential impacts on ingredient availability and costs.

Task: Using Cross-SWOT analysis, develop a summer menu plan that:

1. Leverage strengths to seize opportunities.
2. Minimizes threats using strengths.
3. Mitigates weaknesses through opportunities.
4. Addresses weaknesses and threats comprehensively.

4.5.5 Problem 4.5: Innovation Theory

Amazon is a US-based multinational technology company. It is primarily known for its online retail business, offering a vast range of products, including books, electronics, groceries, and clothing. However, Amazon's influence extends far beyond e-commerce. The company has diversified into multiple sectors, including cloud computing (AWS), digital streaming (Prime Video), smart devices (Alexa, Echo), and physical retail (Whole Foods). Amazon's innovative approach, customer-centric focus, and extensive technological capabilities have made it one of the most influential companies globally.

Task: Answer the following questions with reasons:

1. Is Amazon's innovation an example of sustaining innovation or disruptive innovation?
2. Is Amazon a value innovation? Does it follow Blue Ocean Strategy?

References

Aaker, D. A. (2003). *Strategic marketing management*. Wiley.

Aguilar, F. J. (1967). *Scanning the business environment*. MacMillan.

Blokdyk, G. (2020). *Customer journey map: A complete guide* - 2020 Edition. Emereo Publishing.

Brown, T. (2008). Design thinking. *Harvard Business Review, 86*(6), 84–92.

Buzan, T., & Buzan, B. (1996). *The mind map book: How to use radiant thinking to maximize your brain's untapped potential*. Plume.

Christensen, C. M. (1997). *The innovator's dilemma: When new technologies cause great firms to fail*. Harvard Business School Press.

de Bono, E. (1985). *Six thinking hats: An essential approach to business management*. Little Brown & Company.

Deming, W. E. (1982). *Out of crisis*. MIT Press.

Drucker, P. F. (1954) *The practice of management*. HarperCollins Publishers (2006 version by Harper Business).

Drucker, P. F. (1973). *Management: Tasks, responsibilities, practices*. HarperCollins Publishers (2011 version by Routledge).

Drucker, P. F. (1985). *Innovation and entrepreneurship*. HarperCollins Publishers.

Eberle, B. (2008). *Scamper: Creative games and activities for imagination development* (Combined ed., Grades 2–8). Routledge.

Gray, D., Brown, S., & Macanufo, J. (2010). *Gamestorming: A playbook for innovators, rulebreakers, and changemakers*. O'Reilly.

Harry, M., & Schroeder, R. (2000). *Six sigma: the breakthrough management strategy revolutionizing the world's top corporations*. Doubleday.

Johnson, G., Scholes, K., & Whittington, R. (2008). *Exploring corporate strategy* (8th ed.). Pearson.

Kawakita, J. (1967). *Way of thinking: For creativity development*. Chukoshinsho (in Japanese).

Kim, W. C., & Mauborgne, R. A. (2005). *Blue ocean strategy: How to create uncontested market space and the competition irrelevant*. Harvard Business Review Press.

Kotler, P., Armstrong, G., & Balasubramanian, S. (2023). *Principles of marketing,* Global Edition. Pearson Education.

Kotler, P., Keller, K., & Chernev, A. (2021). *Marketing management*, Global Edition. Pearson.

Learned, E. P., Christensen, C. R., Andrews, K. R., & Guth, W. D. (1965). *Business policy: Text and cases*. Irwin.

McCarthy, E. J. (1960). *Basic marketing: A managerial approach*. Irwin.

Moeran, B. (2005). *The business of ethnography: Strategic exchanges*. Routledge.

Nonaka, I., & Takeuchi, H. (1995). *The knowledge-creating company: How Japanese companies create the dynamics of innovation*. Oxford University Press.

Ohmae, K. (1982). *The mind of the strategist: The art of Japanese business*. McGraw-Hill.

Osborn, A. (1953). *Applied imagination: Principles and procedures of creative problem solving*. Charles Scribner's Sons.

Osterwalder, A., & Pigneur, Y. (2010). *Business model generation: A handbook for visionaries, game changers, and challengers.* Wiley.
Pande, P. S., Neuman, R. P., & Cavanagh, R. R. (2001). *The six sigma way team fieldbook: An implementation guide for project improvement teams.* McGraw-Hill.
Polanyi, M. (1958). *Personal knowledge: Towards a post-critical philosophy.* University of Chicago Press.
Polanyi, M. (1966). *The tacit dimension.* Routledge & Kegan Paul.
Porter, M. E. (1979). How competitive forces shape strategy. *Harvard Business Review, 57*(2), 137–145.
Porter, M. E. (1985). *Competitive advantage: Creating and sustaining superior performance.* Free Press.
Reese, J., Gerwin, K., Waage, M., & Koch, S. (2016). *Value chain analysis: Conceptual framework and simulation experiments.* Nomos.
Ries, A., & Trout, J. (1981). *Positioning: The battle for your mind.* McGraw-Hill.
Schumpeter, J. A. (1912). *Theorie der wirtschaftlichen Entwicklung.* (1983 version: *The Theory of Economic Development: An Inquiry Into Profits, Capital, Credit, Interest, and the Business Cycle.* Transaction Publishers.)
Schumpeter, J. A. (1942). *Capitalism.* Routledge.
Spradley, J. P. (1980). *Participant observation.* Holt.
Weihrich, H. (1982). The TOWS matrix—A tool for situational analysis. *Long Range Planning, 15*(2), 54–66.
Wierzbicki, A. P., & Nakamori, Y. (2006). *Creative space: Models of creative processes for the knowledge civilization age.* Springer.

Further Reading on Knowledge Management and Innovation

Brown, J. S., & Duguid, P. (2017). *The social life of information: Updated, with a New Preface.* Harvard Business Review Press. This book explores how knowledge is created and shared within communities of practice, emphasizing its social and cultural dimensions.
Christensen, C. M. (2024). *The innovator's dilemma, with a new foreword: When new technologies cause great firms to fail.* Harvard Business Review Press. A foundational work that examines why successful companies often fail to adapt to disruptive innovation, offering key insights into the importance of embracing change.
Dalkir, K. (2023). *Knowledge management in theory and practice* (4th ed.). The MIT Press. An essential guide for students, this book provides a solid foundation in the principles and practices of knowledge management.
Davenport, T., & Prusak, L. (2000). *Working knowledge: How organizations manage what they know.* Harvard Business Review Press. Despite its age, this classic text remains a cornerstone in the field, introducing key concepts in knowledge management.
Dyer, J., Gregersen, H., & Christensen, C. M. (2011). *The innovator's DNA: Mastering the five skills of disruptive innovators.* Harvard Business Review Press. Focused on individual innovation, this book identifies five critical skills for disruptive thinking, providing valuable insights for aspiring innovators.
Patriotta, G. (2004). *Organizational knowledge in the making: How firms create, use, and institutionalize knowledge.* Oxford University. Taking a more academic approach, this book delves into the processes of knowledge creation and institutionalization, appealing to those interested in deeper theoretical exploration.

Chapter 5
Decision-Making with the Synthetic Mind

Abstract This chapter explores synthetic thinking and its crucial role in achieving balanced and well-reasoned decision-making. Sensible decision-making integrates rational analysis with intuitive insights, leveraging social information to achieve a practical synthesis. The chapter introduces a comprehensive methodology for knowledge synthesis based on the Knowledge Construction Model and its key component, the Knowledge Triad Synthesis Framework. This framework harmonizes diverse forms of knowledge, enabling well-informed and thoughtful decisions while also facilitating the creation of compelling narratives for effective communication.

5.1 Synthetic Thinking

This section argues that systemic and creative thinking, although seemingly opposed, can complement each other to enhance decision-making. The discussion then shifts from the interaction of different types of thinking to the interplay of different types of knowledge—specifically, how rational and intuitive knowledge can complement each other. Based on this idea, the section defines synthetic thinking and introduces the framework for *sensible decision-making,* which embodies the synthetic decision-making approach advocated in this book.

5.1.1 Systemic and Creative Thinking

Chapters 3 and 4 introduced two crucial approaches: systemic thinking and creative thinking. These approaches are surprisingly complementary in navigating the complexities of the modern world.

Y. Nakamori, *Sensible Decision-Making*, https://doi.org/10.1007/978-981-96-6811-3_5

5.1.1.1 Systemic Thinking

At its core, systemic thinking perceives problems as intricate networks of interconnected elements. It emphasizes:

- *Wholeness*: Recognizing that a system is greater than the sum of its parts.
- *Dynamics*: Understanding how systems evolve and adapt to internal and external pressures.
- *Causality*: Identifying not just immediate causes but also the underlying factors driving an issue.
- *Feedback Loops*: Analyzing how actions within the system influence and are influenced by other elements.

This holistic perspective offers several key advantages:

- Effective Problem-Solving: Addressing the root causes of issues rather than just surface symptoms.
- Long-Term Vision: Fostering sustainable solutions and considering the system's long-term implications.
- System-Wide Optimization: Enhancing overall performance and efficiency rather than focusing on isolated improvements.
- Increased Adaptability: Enabling the system to respond effectively to changing conditions and unforeseen challenges.

Applications of systemic thinking are diverse and impactful:

- Business: Analyzing interdepartmental dependencies, supply chains, and market dynamics for strategic advantage.
- Social Policy: Understanding the root causes of social issues such as poverty, inequality, and crime to develop effective interventions.
- Environmental Management: Developing sustainable practices and policies to address environmental challenges like climate change and resource depletion.

Cultivating systemic thinking enables us to gain deeper insights into complex issues, respond calmly and thoughtfully to challenges, and foster a holistic, interconnected worldview.

5.1.1.2 Creative Thinking

Creative thinking is the engine of innovation, driving us to generate novel ideas and solutions. It emphasizes:

- *Flexibility*: Examining problems from multiple perspectives and experimenting with unconventional approaches.
- *Fluidity*: Shifting thought patterns effortlessly to explore new possibilities and adapt to changing circumstances.

- *Originality*: Producing unique and unexpected ideas that challenge existing assumptions and paradigms.
- *Association*: Linking seemingly unrelated concepts to form new and insightful connections.

This dynamic process offers several key benefits:

- Innovation: Advancing progress across various fields by fostering the development of new products, services, and technologies.
- Effective Problem-Solving: Generating creative and practical solutions to complex, multifaceted challenges.
- Personal Growth: Cultivating curiosity, imagination, and a lifelong passion for learning.
- Enhanced Adaptability: Empowering individuals and organizations to navigate uncertainty and thrive in rapidly evolving environments.

Creative thinking has diverse and far-reaching applications:

- Arts: Inspiring new forms of artistic expression in painting, music, literature, and performing arts.
- Science: Fueling scientific discovery and technological breakthroughs through innovative research and experimentation.
- Business: Driving innovation in product development, marketing strategies, and organizational design.
- Everyday life: Strengthening problem-solving skills, enhancing communication, and enriching both personal and professional experiences.

Developing creative thinking enables us to nurture curiosity, embrace experimentation, and foster a playful, open-minded approach to problem-solving.

5.1.1.3 The Synergy of Systemic and Creative Thinking

Systemic thinking and creative thinking effectively complement each other in decision-making. Systemic thinking offers a framework for understanding interconnections within a system, while creative thinking generates novel ideas and solutions within it.

Together, these approaches provide several key benefits:

- Identifying Root Causes and Generating Solutions: Systemic thinking pinpoints the root cause of a problem, while creative thinking explores innovative solutions.
- Comprehensive Problem-Solving: Systemic thinking examines intricate relationships, while creative thinking introduces fresh perspectives and challenges assumptions.
- Implementing Ideas: Creative thinking generates ideas, while systemic thinking evaluates their feasibility, impact, and potential unintended consequences.

- Strategic Decision-Making: Systemic thinking anticipates system dynamics and future scenarios, while creative thinking envisions possibilities and explores alternative pathways.

This synergistic approach applies to various contexts:

- Problem Structuring: Strengthening systemic thinking through visual tools like system maps to clarify relationships, components, and boundaries over time.
- Brainstorming: Encouraging free-flowing idea generation, fostering creativity, and exploring unconventional solutions.
- Scenario Analysis: Combining systemic and creative thinking to envision and assess future scenarios, potential impacts, and contingency plans.
- Prototyping: Iterating through cycles of systemic analysis and creative refinement to test and improve solutions.
- Feedback Integration: Actively incorporating stakeholder feedback to refine thinking, uncover blind spots, and enhance the approach.

Practical applications span various fields, including:

- New Product Development: Analyzing market trends, customer needs, and competitive landscapes (systemic) to generate and refine innovative product concepts (creative).
- Organizational Transformation: Examining structures, cultures, and processes (systemic) to design and implement new frameworks and operational models (creative).
- Addressing Social Issues: Identifying the root causes of challenges such as poverty, inequality, and environmental degradation (systemic) while developing innovative, sustainable solutions (creative).

Integrating systemic and creative thinking enables us to tackle complex challenges, develop sustainable solutions, and contribute to a better future.

5.1.1.4 Understanding Synthetic Thinking

Humans often engage in synthetic thinking naturally, without consciously using frameworks like systemic or creative thinking. Let us clarify what synthetic thinking is.

Synthetic thinking is a higher-order cognitive skill that synthesizes diverse knowledge—scientific data, cultural perspectives, artistic expressions, and personal experiences—to generate novel insights and a comprehensive understanding. Key aspects include:

- Integration: Actively combining and connecting different types of knowledge, data, and ideas in new and meaningful ways.
- Diverse Knowledge: Utilizing various sources, from quantitative and qualitative data to narratives and lived experiences.

- Novel Insights: Creating new knowledge, identifying previously unseen patterns, and developing innovative solutions that go beyond the sum of their parts.

In today's information-rich world, where data is constantly being generated, synthetic thinking is essential for navigating complexity, making informed decisions, and driving innovation. It empowers individuals and organizations to effectively harness vast amounts of available knowledge to tackle complex challenges.

Distinction from Holistic Thinking: While closely related, synthetic thinking differs from holistic thinking in key ways:

- *Holistic Thinking*: Takes a broad perspective, emphasizing the interconnectedness of all parts within a system and their interactions within the larger context. It provides a framework for understanding the big picture.
- *Synthetic Thinking*: A more specific process that creatively combines diverse knowledge within a holistic framework to generate novel insights and solutions.

Holistic thinking provides the big-picture perspective, while synthetic thinking works within this framework to create innovative solutions. Both holistic and synthetic thinking are valuable cognitive skills that enhance critical thinking, problem-solving, and innovation.

Toward a Synthetic Approach to Decision-Making: Due to human cognitive limitations, mastering holistic thinking can be challenging, even when consciously integrating systemic and creative thinking. Therefore, the following section proposes a decision-making approach that focuses on synthesizing the results of systemic and creative thinking rather than consciously integrating the two thought processes. It suggests an approach that synthesizes rational knowledge, primarily derived from systemic thinking, and intuitive knowledge, mainly generated by creative thinking.

5.1.2 A Framework for Sensible Decision-Making

Effective decision-making requires a diverse set of approaches. These methods can generally be categorized based on their primary mode of reasoning:

1. *Rational Approaches*: These methods prioritize logic, analysis, and evidence-based reasoning. They emphasize objectivity and seek to minimize the influence of emotions or personal biases. Examples include:

    - *Systemic Thinking*: Examines how elements within a system interact, considering long-term consequences and ripple effects.
    - *Critical Thinking*: Focuses on objective analysis, identifying biases, and evaluating evidence to draw well-reasoned conclusions.
    - *Rational Thinking*: Utilizes logic, structured reasoning, and data-driven insights to make informed decisions.

2. *Intuitive Approaches*: These methods rely on instinct, experience, and holistic understanding. They enable quick, adaptive decision-making, particularly in complex or uncertain situations. Examples include:
   - *Creative Thinking:* Generates novel ideas by challenging assumptions and exploring unconventional perspectives.
   - *Lateral Thinking*: Encourages viewing problems from different angles to uncover innovative solutions.
   - *Intuitive Thinking*: Draws on past experiences and subconscious insights to facilitate rapid decision-making, especially under uncertainty.

Balanced Decision-Making: While these categories provide a helpful framework, they are not mutually exclusive. Effective decision-making often integrates rational and intuitive approaches, leveraging their strengths.

5.1.2.1 The Complementary Nature of Rational and Intuitive Knowledge

To understand how rational and intuitive knowledge complement each other, we must examine their characteristics, strengths, and limitations. Table 5.1 provides a comparative overview.

Table 5.1 Key differences between rational and intuitive knowledge

Feature	Rational knowledge	Intuitive knowledge
Characteristics	Logical thinking, data-driven analysis, deductive reasoning, conscious awareness, reliance on facts and evidence	Experience-based judgment, pattern recognition, holistic understanding, tacit knowledge, unconscious awareness
Strengths	Accuracy, objectivity, reproducibility; well-suited for systematic problem-solving, excels in precise calculations and structured analysis	Flexibility, speed, creativity; strong in novel, complex, and ambiguous situations, draws from deep-seated knowledge and experience
Weaknesses	Can be slow, rigid, and overly reliant on data; may overlook contextual nuances and crucial insights	Can be subjective, prone to biases, difficult to articulate, and not always reliable
Examples	Scientific research, financial analysis, legal reasoning, engineering design	Artistic expression, expert judgment in medicine and sports, social navigation, rapid decision-making in emergencies
Role in decision-making	Provides a structured, objective foundation; identifies risks and benefits, ensures consistency and fairness	Offers insights overlooked by pure analysis; enables quick action in time-sensitive situations; fosters creative problem-solving and innovation

Effective decision-making hinges on the interplay between rational and intuitive knowledge. Their complementary roles can be seen in the following ways:

1. Rational knowledge establishes a solid foundation for intuitive judgments, while intuition introduces flexibility and adaptability to rational thinking.
2. Integrating both enhances decision-making by combining objective analysis with creative insights.
3. This integrated approach fosters a deeper understanding of complex problems and enables the discovery of practical solutions.

Examples of complementary use include:

- Scientists apply rational methods for hypothesis testing and data analysis while relying on intuition to identify promising research directions, interpret unexpected results, and recognize subtle patterns.
- Business leaders leverage market research and data analysis to develop strategic plans while using intuition to anticipate market shifts, understand customer reactions, and make swift decisions in uncertain environments.

Rational and intuitive knowledge are not opposing forces but complementary tools. They are best utilized together to enhance decision-making in various fields. Sensible decision-making requires a balanced application of both, blending structure with adaptability to achieve optimal outcomes.

5.1.2.2 Sensible Decision-Making

Sensible decision-making is a general approach that integrates rational and intuitive knowledge, mediated by social knowledge, to support informed and effective decisions. One concrete implementation of this approach is the Knowledge Triad Synthesis Framework, which follows a three-step process:

1. Knowledge Acquisition:
   - Collect Rational Knowledge: Collect relevant data, conduct systematic analyses, and apply logical reasoning to establish a factual, evidence-based foundation.
   - Gather Intuitive Knowledge: Explore diverse perspectives, use creative techniques, and draw from personal and collective intuition to uncover insights, hunches, and unconventional ideas.
   - Gather Social Knowledge: Conduct a stakeholder analysis, considering the values, beliefs, and perspectives of those affected by the decision.
2. Knowledge Refinement:
   - Refine Rational Knowledge: Evaluate rational insights within social and ethical contexts to ensure alignment with societal goals and constraints.
   - Refine Intuitive Knowledge: Assess intuitive insights for cultural feasibility, social acceptability, and ethical integrity.

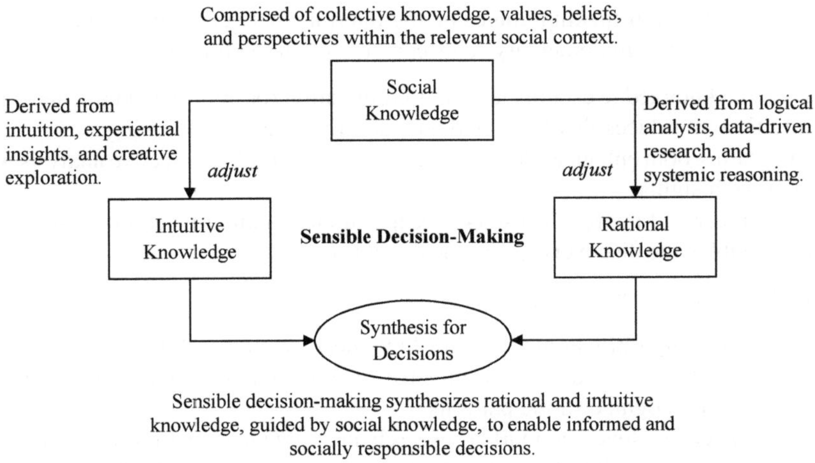

Fig. 5.1 Knowledge synthesis in sensible decision-making

3. Decision Proposal Generation:
   - Synthesize Knowledge: Integrate refined rational and intuitive knowledge to develop well-rounded decision proposals that are:
     a. *Systemically Sound*: Based on a deep understanding of the system and its dynamics.
     b. *Culturally Sensitive*: Aligned with the values and perspectives of affected communities.
     c. *Ethically Responsible*: Mindful of long-term impacts and consequences.

Figure 5.1 illustrates the concept of knowledge synthesis within the context of sensible decision-making. This structured approach is hereafter referred to as the Knowledge Triad Synthesis Framework. *While Sensible Decision-Making is a broad concept, the Knowledge Triad Synthesis Framework provides a concrete method for its implementation.*

Before introducing this framework, let us first examine how researchers and practitioners commonly integrate these three dimensions of knowledge in problem-solving and decision-making.

5.2 Knowledge Triad Approach

This section establishes the foundation for the core theme of this chapter—the *Knowledge Triad Synthesis Framework*, which integrates three key forms of knowledge: *rational, intuitive, and social*—to support sensible decision-making.

The first half of this section examines how prominent thinkers and practitioners have successfully applied this *Knowledge Triad Approach* in their work. The second half explores a specific methodology, demonstrating how the interplay of these three knowledge types can effectively address complex problems.

5.2.1 *Knowledge Triad Models*

A fundamental principle of this chapter is that novel knowledge can emerge by integrating insights from three domains:

1. *Scientific-Actual Domain*
   - Encompasses rational knowledge grounded in evidence and reason.
   - Includes scientific research, logical reasoning, data analysis, and model-based optimization.
2. *Cognitive-Mental Domain*
   - Encompasses intuitive knowledge, such as personal judgments, insights, and beliefs.
   - Includes reasoning based on empirical experience, unique concepts, hypotheses, motivations, and aspirations.
3. *Social-Relational Domain*
   - Encompasses social knowledge distributed within society.
   - Includes social reputation, emerging trends, diverse values, social norms, cultural beliefs, and traditions.
   - Contains both rational and intuitive elements. For example:
     - Established social norms often stem from collective reasoning and shared rational knowledge.
     - Culturally rooted beliefs are typically shaped by shared experiences and intuitive wisdom.

Many researchers have explored problem-solving approaches incorporating similar three-domain models, though these may not fully align with the knowledge domains described above. The following examines some of these alternative perspectives.

Before introducing these models, consider the following clarifications:

1. The three nodes in the knowledge triad models presented below may represent:
   - Knowledge itself
   - Actions that use or create knowledge
   - Domains where knowledge is acquired, or actions are practiced
2. The labels (Rational), (Intuitive), or (Social) beside a node in each model indicate the primary knowledge type associated with that node.
   - These labels highlight how researchers incorporate rational, intuitive, and social knowledge in their work, whether consciously or not.
3. Section 5.3 introduces a model that explicitly distinguishes between knowledge, actions, and domains.
   - It represents actions as nodes and indicates the dominant knowledge type(s) within the surrounding domains.

5.2.1.1 Knowledge Triad Model 1: Schumpeter's Theory of Innovation

Schumpeter's innovation theory exemplifies the Knowledge Triad Approach. As illustrated in Fig. 5.2, successful innovation requires overcoming key barriers that hinder progress.

Economist Joseph Schumpeter (1912) defined innovation as the introduction of novel combinations of production factors—such as resources, labor, and technology—into economic activities. However, innovation faces several barriers (Fig. 5.2), which can be addressed through the following strategies:

- *Existing concepts*: A strong attachment to established paradigms, conventional wisdom, and standard practices can stifle innovation. Organizations must cultivate a culture that actively explores and embraces new ideas.
- *Unexplored territory*: The uncertainty of venturing into uncharted areas can be daunting. Encouraging experimentation and data-driven decision-making enables organizations to navigate these challenges effectively.
- *Social resistance*: Innovation often faces opposition from vested interests and societal norms. Proactively addressing resistance, clearly communicating benefits, and fostering broad support are essential for overcoming this barrier.

Schumpeter's Innovation Theory: Schumpeter placed innovation at the core of economic development and capitalism, emphasizing the role of entrepreneurial activity in driving progress. His key contributions include:

1. The Engine of Change: Schumpeter argued that innovation is the primary driver of economic growth. While established companies may become complacent over time, entrepreneurs disrupt markets by introducing breakthrough innovations.
2. Creative Destruction: Schumpeter (1942), in *Capitalism, Socialism, and Democracy*, introduced the concept of creative destruction, highlighting the dual nature of innovation:

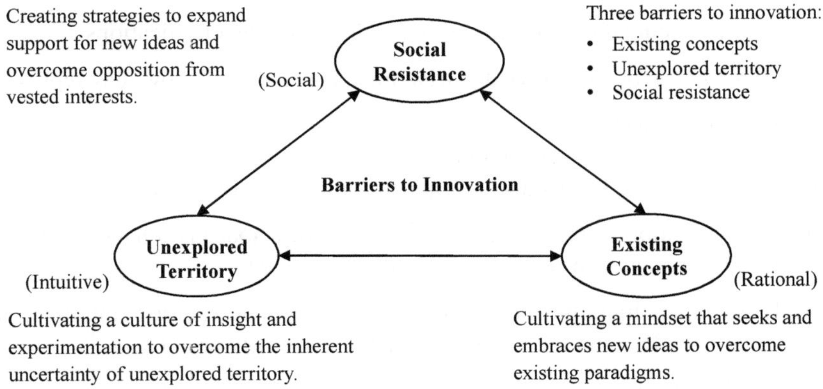

Fig. 5.2 Barriers to innovation and strategies for overcoming them

- It fosters new industries and opportunities.
- It disrupts existing industries, rendering older technologies and business models obsolete.
- This cycle of renewal and disruption fuels economic dynamism, often leading once-dominant firms to decline.

3. The Innovation Process: Schumpeter distinguished between invention (creating new ideas) and innovation (successfully implementing those ideas). Entrepreneurs play a critical role in translating inventions into real-world innovations through key steps such as:
   - Securing funding
   - Assembling a skilled team
   - Navigating market complexities through strategic planning and risk management

Schumpeter's Lasting Influence: Schumpeter's insights continue to shape modern perspectives on economic growth, entrepreneurship, and innovation. His work underscores the importance of fostering an environment that encourages continuous innovation—a critical factor in both business success and broader economic development.

5.2.1.2 Knowledge Triad Model 2: Drucker's Principles for Business Success

Peter Drucker (1954, 1973, 1985) famously stated that a business's primary purpose is to create and retain *customers*. He further argued that companies have only two core functions: *marketing and innovation*. This concept can be visualized as a triad, as illustrated in Fig. 5.3.

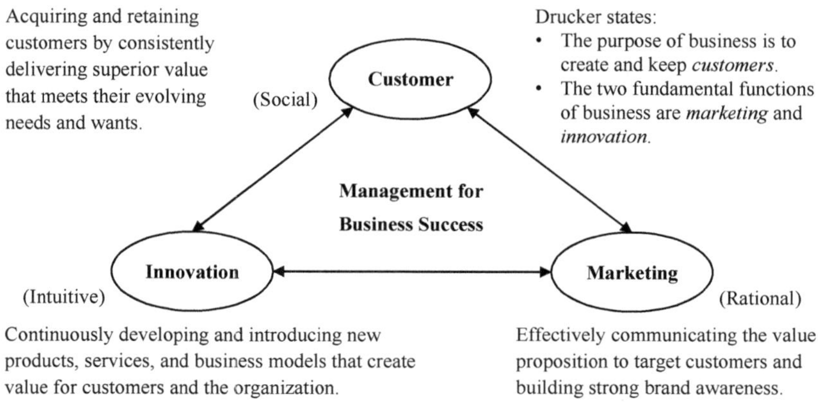

Fig. 5.3 Drucker's business success model

Drucker's work underscores a customer-centric approach to business success, emphasizing three key principles:

1. *Customer Creation as the Prime Directive*: Drucker challenged the traditional view that profit maximization is a business's primary goal. Instead, he argued that:
   - The fundamental purpose of a business is to create and retain customers by consistently delivering value.
   - Meeting customer needs and desires should be the primary and overarching objective.
   - Profitability is not the goal but a natural consequence of effectively fulfilling this core purpose.
2. *Marketing and Innovation as the Two Pillars*: Drucker identified marketing and innovation as the driving forces behind customer creation:
   - Marketing: Understanding & Communicating Value
     - Extends beyond advertising and promotion.
     - Involves deep market research and customer segmentation.
     - Aligns product development with customer needs.
     - Build strong customer relationships through clear and compelling value propositions.
     - Ensures that the entire business adapts to evolving customer expectations.
   - Innovation: Driving Continuous Improvement
     - Encompasses more than breakthrough products.
     - Includes consistent refinement of existing products, services, and processes.
     - Explores new ways to enhance efficiency and deliver greater value.
     - Maintains adaptability in an ever-evolving market.
3. Key Considerations:
   - Supporting Functions: While recognizing the importance of functions like finance, human resources, and production, Drucker emphasized that these should support the core pillars of marketing and innovation.
   - Effectiveness Over Activity:
     - Engaging in marketing and innovation alone is not enough—what matters is effective execution.
     - Success requires translating strategies into real customer value that sustains long-term growth.

Drucker's Lasting Impact: Drucker's core message is clear: Businesses that prioritize customer creation and foster a culture of continuous improvement in marketing and innovation are the ones most likely to thrive in the long term.

5.2.1.3 Knowledge Triad Model 3: Innovation Through Design Thinking

Tim Brown (2008) introduced design thinking as a business approach that applies designers' sensibilities and methodologies to address complex challenges. This approach seeks to:

- Align human needs with technically feasible solutions.
- Translate business strategies into tangible customer value and market opportunities.

This model can be visualized as a triad, as illustrated in Fig. 5.4.

Tim Brown's design thinking framework revolves around balancing three key elements:

1. *Human Desirability*: Understanding Needs & Aspirations
   - Focuses on people first—identifying unmet needs and pain points.
   - Utilizes user research, empathy exercises, and observation to ensure that solutions are meaningful and valuable.
2. *Technological Feasibility*: Navigating Real-World Constraints
   - Evaluates whether proposed solutions are realistically achievable using current technology.
   - Ensures that innovation remains both desirable and technically possible within practical constraints.
3. *Business Viability*: Ensuring Market & Financial Success
   - Assesses the commercial potential of an idea, including profitability and market demand.
   - Ensures that solutions are sustainable and competitive in the long run.

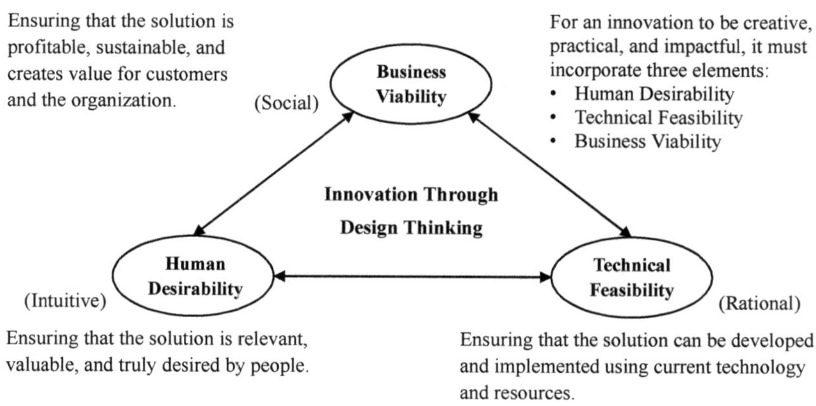

Fig. 5.4 Innovation through design thinking

Tim Brown advocates using designers' unique problem-solving mindset to effectively connect desirability, feasibility, and viability. Key principles include:

a. Empathy & User-Centricity

- Designers excel at understanding user behaviors, needs, and motivations.
- Design thinking ensures that solutions are deeply rooted in human experience.

b. Creative Problem-Solving & Rapid Prototyping

- Encourages exploring multiple solutions instead of relying on conventional approaches.
- Rapid prototyping allows teams to test, iterate, and refine solutions early.

c. Continuous Iteration & Adaptability

- Design thinking is an iterative process—solutions are constantly evaluated and improved.
- This ensures that solutions stay relevant, innovative, and adaptable to evolving market conditions.

Tim Brown argues that design thinking transforms business innovation by ensuring that solutions are:

- Desirable—Aligned with human needs.
- Feasible—Technically achievable.
- Viable—Commercially sustainable.

By integrating these three dimensions, businesses can use design thinking to develop innovative, user-centered, and commercially viable solutions in a constantly evolving market.

5.2.1.4 Knowledge Triad Model 4: System Dynamics and Thought Experiments

Jay Wright Forrester's work on system dynamics (Forrester, 1961, 1969, 1971) introduced a computer-based modeling approach for analyzing complex systems. This approach can be represented as a triad (Fig. 5.5), with causal loop diagrams—shown in the lower left—serving as a key tool for understanding system behavior, especially in innovation.

System dynamics is a powerful framework for analyzing complex systems characterized by feedback loops and nonlinear behavior. It allows researchers to conduct *thought experiments* by simulating system behavior over time and assessing the consequences of various interventions.

Key components of system dynamics are:

1. Causal Loop Diagrams

- Illustrate cause-and-effect relationships among system elements.

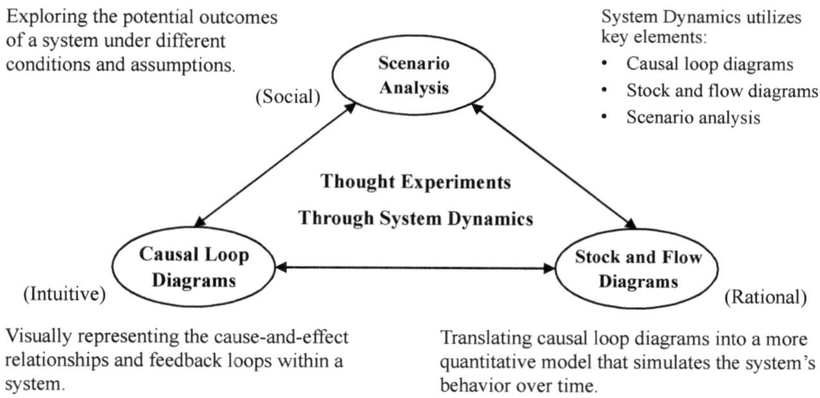

Fig. 5.5 Thought experiments through system dynamics

- Reinforcing loops (+) amplify growth (e.g., innovation driving further innovation).
- Balancing loops (−) stabilize or counteract change (e.g., resource depletion slowing production).
- Identifying these feedback loops is essential for understanding system behavior.

2. Stock and Flow Diagrams
   - Extend causal loop diagrams into quantitative models.
   - Stocks represent accumulations (e.g., knowledge, inventory, population).
   - Flows represent rates of change (e.g., learning rates, production output, birth rates).
   - These models enable researchers to simulate dynamic system behavior using mathematical equations.

3. Scenario Analysis
   - Enables simulated experiments to explore different scenarios.
   - Researchers can analyze potential outcomes, risks, and unintended consequences before adjusting variables or implementing policies in the real world.

Benefits of using system dynamics include:

- Deeper Understanding of Complexity: Breaks down complex systems into manageable components while illustrating intricate interactions.
- Long-Term Perspective: Highlights how present actions shape future consequences, often uncovering hidden risks.
- Informed Decision-Making: Provides policymakers, businesses, and researchers with a data-driven approach to evaluating strategies before implementation.

Applications of system dynamics are diverse and include:

- Business Management: Optimizing supply chains, inventory control, and product development.
- Urban Planning: Simulating transportation systems, infrastructure growth, and resource allocation.
- Environmental Management: Analyzing climate change, pollution, and resource sustainability.

This approach enables researchers to analyze complex systems and conduct thought experiments through:

- Mapping causal relationships through feedback loops.
- Quantifying system behavior with stock and flow models.
- Testing interventions through scenario analysis.

By integrating these methods, system dynamics provides critical insights into decision-making, ensuring that policies and strategies are evidence-based, sustainable, and effective.

5.2.1.5 Knowledge Triad Model 5: Csikszentmihalyi's Systems Model of Creativity

Mihaly Csikszentmihalyi (2015) developed a systems model of creativity that explains how new artifacts emerge within a structured system. His model consists of three elements—the individual, the domain, and the field—interacting to foster creativity. This triadic relationship is illustrated in Fig. 5.6.

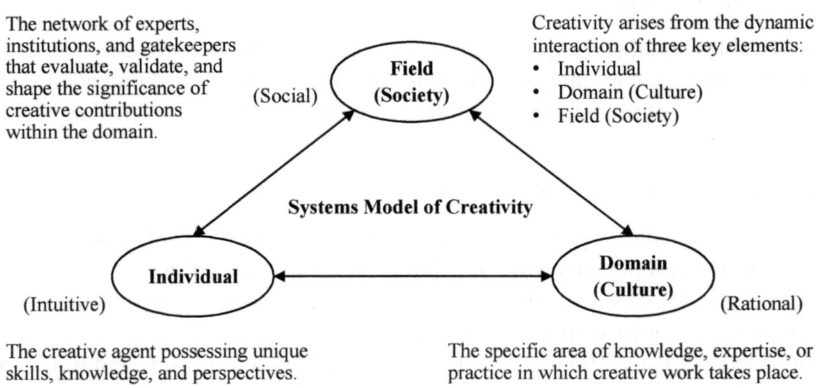

Fig. 5.6 Systems model of creativity

Key elements of creativity are:

1. The Individual
   - The creative agent who possesses unique skills, knowledge, and imagination.
   - Must understand existing domain rules while also challenging conventions and introducing novel perspectives.
2. The Domain
   - The knowledge base of a specific field, such as art, science, or engineering.
   - Comprises established rules, techniques, and frameworks that govern creative work.
3. The Field
   - The community of experts (e.g., art critics, scientific reviewers, industry leaders) responsible for evaluating and validating creative contributions.
   - Serves as a gatekeeper, determining what qualifies as innovation and originality.

Creativity arises from the interplay of these elements:

- The individual generates novel ideas within the domain.
- The field evaluates and validates these ideas.
- This process refines the individual's contributions and shapes the domain over time.

Key insights from Csikszentmihalyi's model include:

- Creativity is Socially Mediated: The field plays a crucial role in recognizing and legitimizing creative work.
- Dynamic Interplay: Creativity does not occur in isolation but emerges from continuous interactions among individuals, knowledge systems, and expert communities.

Criticisms of the model are:

- Overemphasis on Field Validation: Groundbreaking ideas often face initial resistance or rejection by the field, potentially delaying the recognition of radical innovations.
- Underestimation of Individual Genius: The model prioritizes systems and external validation, sometimes overlooking the role of personal talent, motivation, and perseverance in creative breakthroughs.

This model highlights how creativity is shaped by personal ingenuity and external validation, offering a structured view of innovation across fields.

Power of the Knowledge Triad: Knowledge triad models emphasize the importance of examining problems from multiple perspectives to foster creativity and innovation. By integrating three distinct knowledge types—rational, intuitive, and social—individuals can generate unique insights from each perspective.

The true power of this approach lies in identifying connections and overlaps among these domain-specific ideas. These unexpected intersections often spark novel solutions that would not arise from a single, narrow perspective.

Ultimately, knowledge triad models promote a holistic and balanced approach to problem-solving, fostering more creative and resilient solutions.

The following section introduces a systems approach developed in China that incorporates these three knowledge domains. Widely applied in large-scale social projects and corporate management, this approach provides a concrete example of how the framework operates in practice.

5.2.2 A Chinese Systems Approach

This section introduces the WSR systems approach, a framework rooted in Chinese philosophical thought. WSR stands for "Wuli, Shili, Renli," which translates to "Theory of Physics, Theory of Management, and Theory of Humanity." It provides a holistic systems perspective that integrates technical, managerial, and human factors in problem-solving and decision-making.

By the 1970s, Chinese systems scientists recognized the limitations of a purely scientific and technological focus (Wuli). In the late 1980s, Jifa Gu synthesized the concepts of Wuli, Shili, and Renli, highlighting the critical role of human motivation (Renli). Through interactions with systems scientists in the UK, Japan, and China, Gu and his colleagues have refined the WSR systems approach and shaped it into a comprehensive framework.

5.2.2.1 Knowledge Triad Model 6: The WSR Framework

The WSR systems approach (Gu & Zhu, 1995, 2000) provides a structured yet flexible framework for problem-solving and decision-making by integrating diverse perspectives, as illustrated in Fig. 5.7.

The three components of WSR are:

- *Wuli*: Encompasses the physical, material, and technological aspects of a problem, such as natural resources, available technologies, financial constraints, and infrastructure. It is rooted in objective, material-technical realities and relies on technical rationality for analysis.
- *Shili*: Represents the cognitive and methodological dimensions of problem-solving. It includes problem-framing techniques, decision-making methodologies, and analytical approaches. Shili emphasizes pragmatic rationality, focusing on practical and innovative solutions.

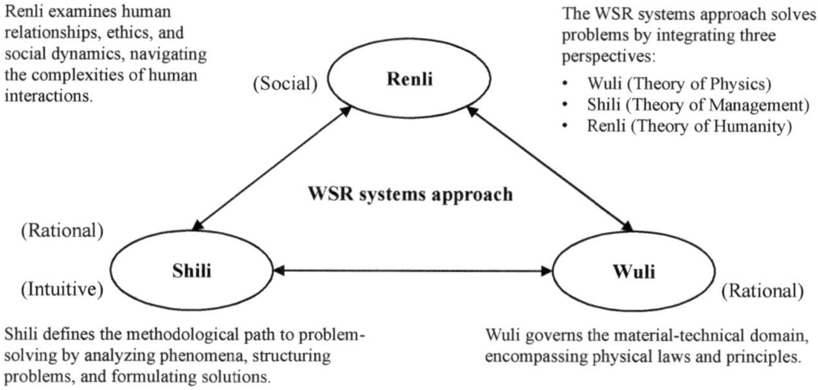

Fig. 5.7 The WSR systems approach

- *Renli*: Addresses the human and social dimensions, including ethical considerations, stakeholder perspectives, cultural influences, and societal impact. It prioritizes ethical rationality, social harmony, and decision legitimacy within a cultural and human context.

The key principles of WSR are:

- *Multi-Perspectival Analysis*: WSR recognizes the limitations of single-perspective approaches and encourages integrating Wuli, Shili, and Renli to understand problems comprehensively.
- *Dynamic Interaction*: Real-world problems are complex and evolving. WSR emphasizes flexibility and adaptability, ensuring that strategies can adjust as circumstances change.
- *Holistic Approach*: By balancing technical, managerial, and human considerations, WSR promotes sustainable and equitable solutions that integrate efficiency, practicality, and social responsibility.

5.2.2.2 Methodological Guidelines

While WSR does not prescribe a rigid sequence of steps, it provides key action elements to guide decision-making:

- Understanding Desires: Identifying the needs and aspirations of stakeholders.
- Investigating Conditions: Analyzing the physical, social, and economic context of the problem.
- Formulating Objectives: Defining clear and achievable goals.
- Creating Models: Developing models and simulations to understand system behavior.
- Implementing Proposals: Translating plans into action.

- Evaluating Performance: Monitoring progress and assessing the effectiveness of implemented solutions.
- Coordinating Relations: Fostering collaboration and communication among stakeholders.

These elements serve as a flexible framework rather than a fixed sequence, allowing for adaptability in addressing complex challenges. WSR recognizes that the relative importance of Wuli, Shili, and Renli varies based on context and evolving problem dynamics.

WSR distinguishes between "hard WSR" and "soft WSR," reflecting its dual nature:

- Hard WSR: Used in quantitative and mathematical problem-solving approaches.
- Soft WSR: Emphasizes participatory methods and qualitative considerations, particularly in human-centric decision-making.

As Zhu (2023) highlights, this distinction underscores WSR's integration of both hard (quantitative, data-driven) and soft (qualitative, human-centered) aspects. Figure 5.7 illustrates how Shili bridges rational and intuitive knowledge within this framework. Additionally, Table 5.2 (prepared by Zhu, 2023) provides a detailed comparison of these perspectives. In the table, SE and OR stand for systems engineering and operations research, respectively.

5.2.2.3 A Case Study: Investigating the Meta-Synthesis Systems Approach

Gu and Tang (2004) examine the application of the WSR systems approach in a significant research project on the Meta-Synthesis Systems Approach (MSA). Since the late 1980s, Chinese systems scientists have explored MSA (Qian et al., 1990) to tackle complex systems problems by integrating diverse knowledge sources and methodologies.

This study analyzes how the core elements of WSR—Wuli, Shili, and Renli—were manifested in the MSA research project:

- Wuli: Represents the project's infrastructure, including funding, research facilities, and the expertise of participating institutions.
- Shili: Encompasses the methodological framework, including models, data, and knowledge discovery techniques.
- Renli: Focuses on human dynamics, such as communication, collaboration, and trust among researchers. Social network analysis was applied to study interaction patterns within the research team.

This case study highlights WSR's value as a framework for analyzing complex research projects, particularly in human–machine collaboration and interdisciplinary research. By explicitly incorporating Wuli, Shili, and Renli, researchers can better

Table 5.2 Hard WSR vs. soft WSR

Aspect	Hard WSR	Soft WSR
Positioning WSR	A precise, multi-criteria decision-making tool for SE/OR experts	A multi-perspective, multi-method, stakeholder-driven management approach
Accountability	To authorities and business sponsors	To stakeholders, particularly those affected
Scope of integration	Expert knowledge across disciplines, integration of quantitative and qualitative methods, and human–machine collaboration	Wuli fundamentals, Shili creativity, and Renli human relations—especially stakeholder interests, values, and expectations
Role of non-expert stakeholders	Subjects of study, management, and oversight	Active owners and participants in decision-making
Handling of Renli (human relations)	Navigating relational complexities to achieve project goals	Enhancing ethical awareness by asking, "To what end?"
Scientific vs. humanistic approach	"From qualitative to quantitative"—all human problems should be quantified and analyzed	Opposes reductionism—value rationality requires contextual observation, participation, and interaction
Functionality of SE/OR	Wuli is addressed by natural science; SE/OR (as applied mathematics) forms the foundation of Shili-ology	SE/OR techniques apply technical rationality to Wuli, while their pragmatic rationality supports Shili
Key resources	Databases, models, methods, hardware, software, groupware, harmony-ware, people-ware, heart-ware, etc	Participants' experience, skills, and creativity in setting agendas and designing decision-making strategies
Modes of application	Industry SE, social SE, agriculture SE, defense SE, education SE, etc	Formal projects, informal problem-solving, and adaptive responses

understand the factors driving project success and identify opportunities for improved project management and design.

5.2.2.4 Application: Developing a Water Resources Decision Support System

The WSR approach has been successfully applied to formal management projects, guiding all phases from funding and task assignments to project completion and evaluation. Gu and Tang (2000) present a case study on the development of a decision support system for water resources in a port city in northern China.

The local government faced a severe water shortage, worsened by growing tourism and poor water management. In response, it invested in water infrastructure, including diversion tunnels. However, the newly established Water-Diverting Management Bureau struggled to manage the expanded water supply and balance competing demands. Recognizing the issue's complexity and the Bureau's limited expertise, the government sought assistance from a WSR team of systems scientists.

The WSR team initiated the project by engaging stakeholders in the "understanding desires" phase, uncovering key communication challenges:

- User groups struggled to articulate their water needs.
- Bureau staff were impatient with user concerns.
- City officials had not yet clearly defined their objectives.

To bridge these gaps, the team conducted the "investigating conditions" phase through interviews, stakeholder workshops, and on-site surveys, leading to a nuanced understanding of the system:

- Wuli (Physical Aspects): Identified key water sources (rivers, tunnels, reservoirs, groundwater), seasonal variations, and treatment plant capacities.
- Shili (Methodological Aspects): Revealed conflicting stakeholder priorities:
  - Reservoir operators prioritized stable water levels.
  - Water treatment plant operators focused on workload optimization.
  - Bureau managers favored a market-based allocation.
  - User groups advocated for fair distribution.
  - City officials sought economic, social, and environmental balance.
- Renli (Human Aspects): A win-lose mentality prevailed among competing user groups (industrial, agricultural, domestic, and tourism), with each prioritizing its interests.

Acting as a neutral facilitator, the WSR team guided stakeholders through discussions centered on three key questions:

- What do we want?
- What do we have?
- What should we seek?

This approach helped stakeholders:

- Understand each other's perspectives.
- Identify common ground.
- Explore mutually beneficial solutions.

Over time, mindsets shifted from adversarial to collaborative.

The project employed various analytical tools, chosen based on stakeholder needs, not imposed by external experts:

- Rich pictures: Visualized stakeholder relationships and water flow.
- Cognitive mapping: Captured diverse viewpoints.

- Critical path analysis: Optimized decision-making timelines.
- Input–output tables: Assessed economic and resource trade-offs.

The project faced several obstacles:

- Data gaps in water availability and consumption.
- Integration of multidisciplinary knowledge (engineering, economics, social sciences).
- Complex interpersonal dynamics among government agencies, user groups, and researchers.

Despite these challenges, the project successfully developed a decision support system that has since proven effective in managing water supply during droughts. This case demonstrates the practical value of the WSR approach in tackling complex water management issues.

5.3 An Approach to Knowledge Synthesis

This section presents the knowledge construction methodology that guides:

1. Planning knowledge collection for solving a given problem.
2. Collecting and constructing knowledge within three domains: Scientific-Actual, Cognitive-Mental, and Social-Relational.
3. Integrating obtained domain-specific knowledge to generate broader, more comprehensive insights.

The first half explains the methodology, while the second applies it to the critical issue of regional revitalization.

5.3.1 The Knowledge Construction Methodology

This subsection introduces the core components of the knowledge construction methodology (Nakamori, 2019):

- The knowledge construction model
- The knowledge construction diagram
- The knowledge construction guidelines

To illustrate how knowledge is synthesized through the above domains, we examine expertise development in Go, though similar principles apply to Chess and other board games (see Fig. 5.8).

Go is an ancient Chinese board game renowned for its strategic depth and complexity. Players alternate placing black and white stones on a 19×19 grid, aiming to control territory by surrounding empty spaces. While the rules are simple, mastery

Fig. 5.8 The development of expertise: insights from Go and other board games

requires anticipating long-term consequences, adapting to evolving strategies, and demonstrating tactical finesse.

As shown in Fig. 5.8, professional Go players develop expertise through three interconnected domains of learning:

I. Learning Established Tactics (Scientific-Actual Domain):
   - Building a Strong Foundation: Understanding established tactics enhances creativity in the idea generation phase.
   - Pattern Recognition: Analyzing top players' strategies helps identify applicable tactical patterns, reinforced through practical application in the idea verification phase.

II. Creating New Tactics (Cognitive-Mental Domain):
   - Continuous Ideation: Advanced players must consistently generate innovative strategies beyond established knowledge.
   - Adapting to the Meta-Game: Innovation is crucial in response to evolving gameplay dynamics and opponents' strategies.

III. Verifying Learned or Created Tactics (Social-Relational Domain):
   - Testing Against Strong Opponents: Competing with skilled players pushes limits and fosters improvement.
   - Learning from the Best: Observing and analyzing top players deepens tactical knowledge and inspires innovation.

Mastery in Go requires a combination of structured learning, creative exploration, and real-world application—but dedication, talent, and passion also play a significant role.

Human Learning vs. Machine Learning—The Case of AlphaGo: AlphaGo, developed by Google DeepMind, revolutionized Go using artificial intelligence. Unlike human players, AlphaGo learns primarily through machine learning algorithms rather than human experiences like mentorship, intuition, or emotional engagement.

Question: How do human Go players learn and improve compared to AlphaGo? (End-of-Chapter Problem 5.2)

- Human Players: Learn by studying past strategies, experimenting with innovative ideas, and competing against strong opponents. This process integrates scientific, cognitive, and social knowledge domains.
- AlphaGo: Utilizes deep neural networks and reinforcement learning, analyzing millions of game scenarios to identify optimal moves without relying on human intuition or traditional learning structures.

Unlike human players, who rely on experience, intuition, and social learning, AlphaGo optimizes decisions purely through reinforcement learning. It often makes moves that defy conventional strategic wisdom. By contrasting human learning with AI-based approaches, we gain deeper insights into the nature of knowledge synthesis and adaptation in complex systems.

5.3.1.1 The Knowledge Construction Model (*i*-System)

As mentioned briefly in Sect. 5.2.1, the knowledge construction model represents actions as nodes and highlights the domains in which each piece of knowledge is dominant.

The model, shown in Fig. 5.9, consists of five action nodes, each beginning with the letter "*i*." The upper three nodes focus on constructing knowledge—through collection, organization, or creation—within distinct domains. The lower two nodes guide the work of the upper nodes and synthesize the knowledge they generate. The upper nodes operate within the scientific-actual, social-relational, and cognitive-mental domains, while the lower nodes function within the initiative-creative domain.

Before exploring the knowledge domains and action nodes in detail, it is important to define key terms to distinguish specific processes:

1. Knowledge Construction vs. Knowledge Creation
   - While closely related, these terms are not identical. Knowledge construction refers to the systematic gathering and organization of knowledge, often serving as a foundation for knowledge creation, which involves generating entirely new insights.
   - Both processes rely on creativity and critical thinking, requiring information analysis and well-reasoned judgment.
2. Knowledge Integration vs. Knowledge Synthesis
   - Both involve combining separate pieces of knowledge but differ in scope and purpose. Knowledge integration is assembling relevant knowledge from various sources, laying the groundwork for knowledge synthesis, which generates novel insights by merging and reinterpreting integrated knowledge.

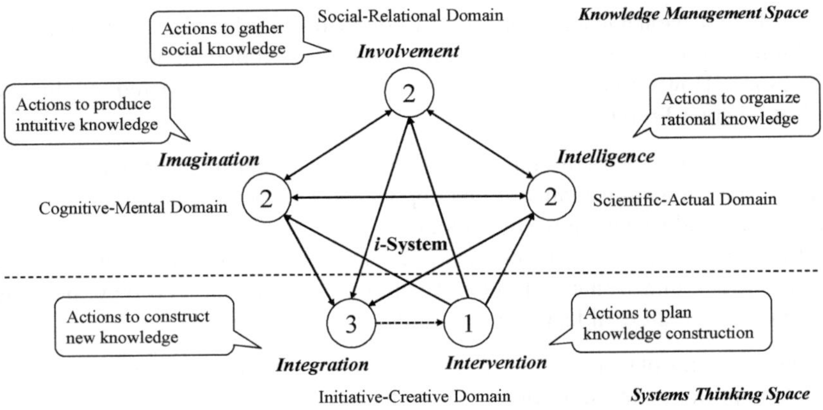

Fig. 5.9 The knowledge construction model

- Knowledge integration is a prerequisite for synthesis; meaningful, innovative ideas cannot emerge without first establishing connections between existing knowledge.

The following explains the knowledge domains and action nodes involved in developing a research plan.

Knowledge Domains: The scientific-actual, social-relational, and cognitive-mental domains are central to knowledge management. While not entirely distinct, they encompass rational, social, and intuitive knowledge. The initiative-creative domain, rooted in systems thinking, plays a unique role: it provides the strategic perspective necessary to set goals, consider the broader context, and guide knowledge construction by recognizing interrelationships among elements. Ultimately, this model aims to integrate systems thinking with knowledge management.

Knowledge across the four domains is organized as follows:

A. **Scientific-Actual Domain**

- This domain includes rational knowledge from empirical evidence, such as science, technology, socio-economic trends, academic assessments, and historical facts.
- In the context of research, this domain includes:
  1. Established knowledge from prior research.
  2. Essential data and information for conducting research.
  3. Knowledge technologies such as mathematical modeling and analytical tools.

B. **Social-Relational Domain**

- This domain includes social knowledge shaped by norms, values, cultures, power dynamics, reputations, traditions, and emerging trends.
- In the context of research, this domain includes:

   1. Ideas from external collaborators within research networks.
   2. Insights from big data and knowledge disseminated in society.
   3. Knowledge technologies such as data mining and text analysis.

C. **Cognitive-Mental Domain**

- This domain includes intuitive knowledge derived from individual judgment, including decision-making criteria, dominant logic, conceptual frameworks, hypotheses, motivations, and aspirations.
- In the context of research, this domain includes:
  1. Experiential knowledge that inspires and motivates research.
  2. Ideas generated by individuals or internal research groups.
  3. Knowledge technologies related to organizational knowledge creation.

D. **Initiative-Creative Domain**

- This domain includes strategic knowledge that facilitates knowledge collection, construction, and evaluation.
- In the context of research, this domain includes:
  1. Methods for developing a research plan.
  2. Strategies for promoting a research project.
  3. Approaches to organizing research results.
  4. Techniques for publicizing research findings.

Action Nodes: Knowledge construction unfolds in three interconnected stages:

I. *Planning*: At the **Intervention** node, a structured plan for knowledge collection and construction is developed. This stage involves:

- Defining problems, goals, and evaluation criteria.
- Establishing a research framework.
- Determining key activities within the three knowledge domains.

II. *Knowledge Collection*: In this stage, domain knowledge is gathered or generated through three key nodes: Intelligence, Involvement, and Imagination. The sequence of these nodes is flexible, allowing for iteration when necessary.

- **Intelligence**: Collecting, organizing, and analyzing existing knowledge and relevant data.
  1. Conducting literature reviews.
  2. Gathering essential data.
  3. Learning analytical and modeling techniques.
- **Involvement**: Acquiring external insights and implicit knowledge through interaction and collaboration.
  1. Identifying and engaging stakeholders.
  2. Collaborating with external researchers.
  3. Securing support from colleagues and institutions.

- **Imagination**: Generating innovative ideas by processing collected knowledge and fostering creative interaction.
    1. Utilizing codified and networked knowledge.
    2. Encouraging innovative thinking.
    3. Integrating personal and collective insights.

III. *Knowledge Integration and Evaluation*: At the **Integration** node, knowledge from the three domains is synthesized and assessed. This stage includes:

- Constructing new knowledge (e.g., research findings).
- Producing academic papers, reports, or books.
- Identifying new research challenges for future exploration.

If the synthesized knowledge does not meet predefined evaluation criteria, the process loops back to earlier stages for refinement and improvement.

Tacit Integration—A Key to Knowledge Construction: A central challenge in the knowledge construction model is the effective implementation of the *Integration* node. To address this, we can draw inspiration from Michael Polanyi's concept of tacit integration.

Polanyi (1958, 1966) argued that much of human knowledge is tacit—intuitive, experiential, and difficult to articulate explicitly. For example, we can effortlessly recognize a familiar face or grasp a complex concept yet struggle to explain the underlying cognitive processes. These examples illustrate that explicit knowledge, which can be expressed in words, numbers, or formulas, represents only a fraction of our understanding. More importantly, they suggest that we often make decisions and generate insights by subconsciously integrating diverse knowledge and experiences.

Polanyi called this ability "*tacit integration*," referring to the subconscious process of synthesizing explicit and tacit knowledge to form a coherent understanding. This ability enables pattern recognition, problem-solving, and creative insight, allowing individuals to draw meaningful conclusions from fragmented information—often without being able to explicitly explain how they arrived at them.

Key aspects of tacit integration include:

- Subconscious Nature: Much of the integration process occurs below the level of conscious awareness, relying on intuition and experience.
- The Inseparability of Explicit and Tacit Knowledge: Explicit knowledge is always rooted in tacit understanding; we cannot fully apply explicit knowledge without its underlying tacit foundation.
- Holistic Understanding: Tacit integration enables us to perceive the "big picture" and recognize connections between seemingly unrelated elements.
- Role of Experience: As we gain experience, we improve our ability to synthesize information and generate insights.

Recognizing the power of tacit integration helps us appreciate the limitations of explicit knowledge and the critical role of tacit knowledge in cognition, learning, and creativity. These insights can guide the design of more effective knowledge construction models and strategies.

The *Integration* node of the knowledge construction model relies heavily on tacit integration. However, to facilitate this process, the knowledge construction methodology provides a framework that systematically supports integration. Before exploring this framework, let us compare this human ability with AI deep learning.

Tacit Integration vs. Deep Learning: Tacit integration is a human cognitive process that relies on intuition, experience, and contextual understanding. In contrast, deep learning is a machine learning technique that utilizes artificial neural networks to recognize patterns and make predictions. While both processes involve pattern recognition and decision-making, they differ fundamentally:

- *Tacit Integration*: Subconscious, holistic, and deeply influenced by lived experience.
- *Deep Learning*: Data-driven, algorithmic, and explainable through mathematical models.

Question: How do human tacit integration and AI-based techniques differ in the context of facial recognition? (End-of-Chapter Problem 5.3)

5.3.1.2 The Knowledge Construction Diagram

The knowledge construction methodology provides a systematic approach to synthesizing knowledge, leveraging the human capacity for tacit integration. Figure 5.10 illustrates this methodology through a diagram that integrates knowledge from three domains, demonstrating how problem-solving ideas emerge.

Begin by populating the three top boxes with your understanding of the following knowledge domains:

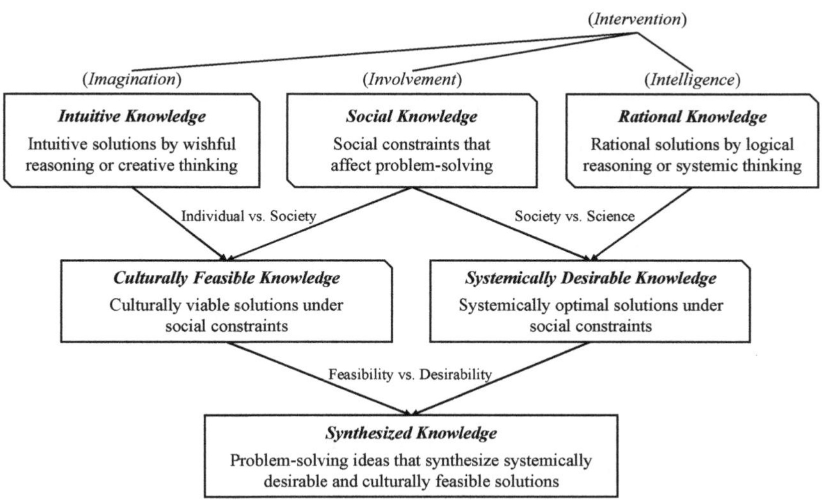

Fig. 5.10 The knowledge construction diagram

- Rational knowledge involves quantitative relationships, such as simulation or optimization results, and should be expressed clearly and concisely.
- Social knowledge encompasses large volumes of textual and numerical data, requiring careful summarization and documentation.
- Intuitive knowledge represents a structured form of tacit knowledge, demanding deep understanding and precise articulation.

The middle section of Fig. 5.10 represents intermediate knowledge, derived as follows:

- Systemically desirable knowledge refines rational knowledge while considering the constraints of social knowledge.
- Culturally feasible knowledge emerges from intuitive knowledge, shaped by the constraints of social knowledge.

Finally, the bottom box represents synthesized knowledge, integrating systemically desirable and culturally feasible ideas to generate problem-solving solutions. Successful integration relies on tacit integration, where intuition plays a key role—but logical consistency must be maintained.

Systemic Desirability and Cultural Feasibility: Systemic desirability and cultural feasibility originate from the Soft Systems Methodology (Checkland, 1981). These interconnected principles are essential for a system's success.

A. *Systemic Desirability*: A system is systemically desirable when it:
   - Fulfills its intended purpose and meets stakeholder needs and expectations.
   - Operates efficiently, optimizing resource utilization.
   - Produces expected outcomes and contributes effectively to problem-solving.
   - Functions as an integrated whole, with well-aligned and cohesive components.
   - Harmonizes with its environment and other systems, ensuring coexistence.

B. *Cultural Feasibility*: A system is culturally feasible when it:
   - Is understood, accepted, and actively used by stakeholders.
   - Aligns with stakeholder values, fostering engagement and trust.
   - Respects existing practices and cultural norms, enabling smooth implementation.
   - Operates without social resistance or unintended negative consequences.
   - Adheres to ethical standards and human rights, preventing misuse.

Balancing Systemic Desirability and Cultural Feasibility: For a system to be effective, both systemic desirability and cultural feasibility must be balanced. A systemically desirable idea that lacks cultural feasibility may face resistance, hindering implementation. Conversely, a culturally feasible idea without systemic desirability may fail to meet stakeholder needs, rendering it ineffective. Achieving both ensures practical and sustainable solutions.

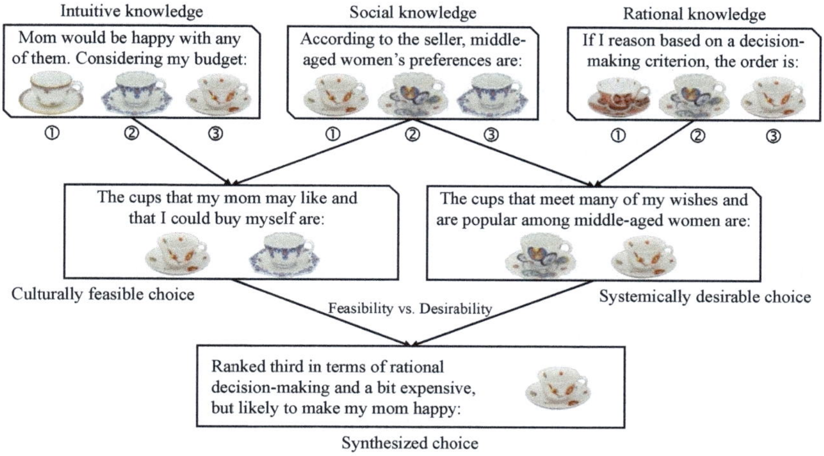

Fig. 5.11 Selection of a coffee cup for one's mother

Example—The Ideal Choice: Fig. 5.11 illustrates the application of the knowledge construction diagram to the problem of selecting a coffee cup for one's mother, as discussed in Chapter 2. While this example achieves a well-integrated solution, real-world decision-making often requires compromises, making such definitive conclusions challenging.

The knowledge construction diagram provides a structured framework for making choices that are both personally satisfying and persuasive to others, even when compromises are necessary.

5.3.1.3 The Knowledge Construction Guidelines

The knowledge construction methodology follows guidelines for effectively collecting, synthesizing, and evaluating knowledge. Initially based on multimedia, emergent, and evolutionary falsification principles (Wierzbicki & Nakamori, 2007), it now incorporates the intervention principle (Nakamori, 2021).

1. Multimedia Principle
   - Focus: Collect and explore knowledge using diverse media (text, visuals, audio, sensory data).
   - Rationale: Relying on a single source limits understanding, whereas multimodal input enhances creativity, comprehension, and a holistic grasp of reality.
   - Application: Actively seek and integrate information from multiple sources to build a richer knowledge base.
2. Intervention Principle

- Focus: Engage proactively in knowledge creation rather than relying solely on passive observation.
- Rationale: Direct intervention fosters deeper understanding through experimentation, model-building, and iterative refinement.
- Application: Formulate hypotheses, conduct experiments, and analyze results to generate new insights.

3. Emergent Principle

- Focus: Recognize that new knowledge emerges from the interaction of existing concepts.
- Rationale: Systematic organization and meaningful connections between ideas lead to novel insights and innovative solutions.
- Application: Encourage interdisciplinary collaboration and explore unexpected connections across different fields.

4. Evolutionary Falsification Principle

- Focus: Continuously refine and validate knowledge through rigorous evaluation and critical dialogue.
- Rationale: Knowledge evolves through testing, refinement, and adaptation rather than remaining static.
- Application: Engage in open, constructive debate, address critiques, and revise understandings based on new evidence.

5.3.2 Application to Business Planning

The knowledge construction methodology shares structural similarities with the WSR systems approach but differs in several key aspects.

The WSR systems approach examines issues from material, management, and social perspectives. It defines problems from a management standpoint and seeks solutions within the constraints of material and social factors. The social perspective is then used to refine these solutions. Alternatively, a social problem-solving approach can be employed, where the three perspectives interact dynamically, adjusting their thinking to reach a compromise.

In contrast, *the knowledge construction methodology identifies optimal solutions within each of the three knowledge domains while minimizing interference from other domains. These domain-specific solutions are then synthesized to derive a sensible, well-balanced solution.*

This subsection explores how to apply this methodology to social project planning. In practice, project implementation requires continuous plan adjustments and ongoing interaction and learning among stakeholders.

Application—Food Recycling Business Planning: This application draws on the experience of a Japanese food recycling business to illustrate how knowledge (ideas) can be constructed using the knowledge construction model in Fig. 5.9.

Imagine a scenario in which policymakers in a regional city aim to address global environmental concerns within their community. They focus on food recycling—a process that involves collecting food waste, composting it, and using the compost to cultivate and sell vegetables.

The business plan is developed through the following stages:

- Stage 1—*Intervention Node*: Define the knowledge (ideas) to be collected and created at the Intelligence, Involvement, and Imagination Nodes.
- Stage 2—*Intelligence, Involvement, and Imagination Nodes*: Collect and generate knowledge (ideas) within each domain.
- Stage 3—*Integration Node*: Integrate and synthesize the three types of knowledge into a cohesive plan.

I. **Stage 1—Planning at the Intervention Node**: At this stage, policymakers define how they will acquire or create knowledge across the three domains:

   - *Intelligence*: Conduct a scientific assessment of the environmental and economic impacts of converting food waste into compost and evaluate its potential benefits.
   - *Involvement*: Understand the perspectives of compost manufacturers and farmers regarding the business—additionally, survey citizens who provide food waste and those who purchase agricultural products.
   - *Imagination*: Explore management strategies to establish the food recycling business as a model for environmentally sustainable industries.

II. **Stage 2—Acquiring and Creating Knowledge at Three Action Nodes**:

   A. **Intelligence**—*Gathering Scientific Knowledge on Environmental and Economic Impacts*: Composting presents a promising alternative to incineration for reducing greenhouse gas emissions. It utilizes microorganisms to break down food waste into nutrient-rich compost. However, both environmental impact and economic feasibility must be carefully considered.

      - Environmental Impact: Simulations suggest that current composting technology may produce slightly higher carbon dioxide emissions than incineration. To mitigate this, advancements in composting technology, the use of electric vehicle collection, and more efficient food waste collection methods are necessary. If these innovations succeed, composting could become a viable business model for reducing greenhouse gases.
      - Economic Viability: For a composting business to be profitable, it must operate at scale, ensuring a sufficient supply of food waste and a stable market for compost sales. However, the initial investment in infrastructure and equipment is substantial, posing a financial challenge in the early stages. Overcoming these hurdles could establish the business as a model for local economic revitalization.

B. **Involvement**—*Understanding Stakeholder Perspectives*: Policymakers are surveying key stakeholders—including compost manufacturers, farmers, and citizens—to assess their attitudes toward the food waste composting initiative. While stakeholders generally support environmental improvements and economic development, their enthusiasm remains lukewarm.

- Compost Manufacturers: To ensure business continuity, they require a stable supply of food waste and guaranteed sales channels for the compost produced. Despite the significant upfront investment, they emphasize the need for innovative composting technologies that minimize greenhouse gas emissions.
- Farmers: Many hesitate to switch from chemical fertilizers—readily available through agricultural cooperatives—to the slightly more expensive compost derived from food waste. They also worry about consumer acceptance of vegetables grown with compost in the market.
- Citizens: Residents express concerns about the inconvenience of drying food waste at home before bringing it to designated collection points. Additionally, they are unsure about the safety of vegetables grown using food-waste-based compost. However, they recognize the importance of environmental conservation and are willing to cooperate if they see tangible environmental benefits.

C. **Imagination**—*Envisioning a Sustainable and Economically Vibrant City*: City policymakers aim to transform the community into an environmentally conscious and economically thriving city through food waste composting.

- Revitalizing the Local Economy: The city is experiencing economic stagnation due to population decline. Policymakers seek to reinvigorate local industries and improve residents' quality of life. Promoting food waste composting as a key industry could catalyze economic growth. While initial investments in composting infrastructure are necessary, future cost savings from reduced food waste incineration could offset these expenditures.
- Creating an Environmentally Friendly City: Policymakers recognize the need to raise awareness among business owners, farmers, and the general public to foster sustainable business practices. They plan to actively promote environmental education through local organizations, such as women's associations, to build understanding and encourage community participation in the composting initiative.

III. **Stage 3—Knowledge Synthesis and Evaluation at the Integration Node**: At this stage, ideas from the three knowledge domains—through the Intelligence, Involvement, and Imagination Nodes—are synthesized to develop a viable and sustainable food waste composting strategy.

1. **Systemically Desirable Ideas**: City authorities must identify composting companies capable of operating at scale while ensuring profitability through a sustainable business model that covers operational costs.

   - Environmental Impact: The selected composting technology should minimize greenhouse gas emissions compared to incineration. Authorities must also evaluate the initial investment and assess its long-term sustainability.
   - Financial Incentives: The city can provide subsidies to offset initial capital costs and support early-stage composting operations. These subsidies can be recovered through long-term savings from reduced incineration expenses.
   - Collaborative Partnerships: Collaboration between city authorities, compost manufacturers, and farmers is crucial. Without a stable market for compost, an oversupply would lead to waste, undermining environmental benefits and necessitating incineration. Securing farmer participation is a top priority.

2. **Culturally Feasible Ideas**: The success of a food waste composting project depends on active citizen participation—both in supplying food waste and in purchasing compost-grown vegetables.

   - Citizen Engagement: A user-friendly food waste collection system is essential. Developing this system involves educating residents on proper waste storage methods and establishing convenient collection points. Community-driven environmental education can foster responsibility and encourage participation.
   - Market Development: The program's long-term success depends on increasing consumer demand for compost-grown vegetables. Public education campaigns should highlight the environmental and potential health benefits.
   - Building an Environmentally Conscious City: Raising public awareness through environmental education initiatives can promote widespread cooperation and support for the composting project.

3. **Synthesized Strategy**: A successful food waste composting initiative must balance environmental responsibility, economic sustainability, and community engagement. Key strategies include:

   - Reducing Environmental Impact: Implementing composting methods that significantly lower greenhouse gas emissions.
   - Ensuring Economic Viability: Establishing a financially sustainable business model, incorporating government subsidies, and securing a substantial market for compost sales.
   - Farmer Cooperation: Organizing a network of farmers committed to purchasing and using compost for vegetable cultivation.

- Encouraging Citizen Participation: Developing an accessible food waste collection system and fostering environmental responsibility through education and community initiatives.
- Promoting Market Development: Encouraging the consumption of compost-grown vegetables through awareness campaigns and farmer engagement.

Conclusion: The success of this food waste composting project depends on the implementation of the synthesized strategies mentioned above. This approach benefits the environment and revitalizes the local economy.

5.4 Decision-Making Through Knowledge Synthesis

The Knowledge Triad Synthesis (KTS) Framework supports the central theme of this book, *sensible decision-making*. This framework, an independent adaptation of the *Integration* stage in the knowledge construction model, offers a structured approach to synthesizing knowledge from diverse sources. The latter half of this section demonstrates how the KTS Framework can be applied to develop a strategy story.

5.4.1 The Knowledge Triad Synthesis Framework

The Knowledge Triad encompasses various dimensions of knowledge, including its creation, sharing, utilization, structure, form, management, and philosophical foundations. This section focuses on three primary types of knowledge essential for decision-making: rational, intuitive, and social. While these types are interconnected, they are categorized here for analytical clarity.

1. Rational knowledge
    - Logical Inference: Knowledge derived through logical reasoning.
    - Data Analysis: Patterns and trends identified through data analysis.
    - Model Analysis: Optimal solutions and simulations generated using mathematical models.
2. Intuitive knowledge
    - Experience-Based Foresight: Insights gained from personal experiences and intuition.
    - Creative Ideas: Novel concepts developed through innovative thinking.
    - Collaborative Ideas: Ideas refined through group discussions and shared knowledge.

3. Social knowledge

- Social Trends: Socioeconomic factors influencing decision-making.
- Social Constraints: Limitations and enabling factors affecting implementation.
- Diverse Values: The varied perspectives and priorities of individuals and groups.

Figure 5.12 illustrates the KTS Framework, aligning with the knowledge construction diagram.

Knowledge Synthesis Procedure: The KTS Framework follows a structured process for acquiring, creating, and synthesizing knowledge. The three stages correspond to the knowledge construction model:

I. **Stage 1—Intervention**:

- Define the Problem: Clearly articulate the decision problem and its context.
- Develop a Knowledge Acquisition Plan: Identify the specific types of knowledge needed from each domain (rational, intuitive, and social).

II. **Stage 2—Intelligence, Imagination, Involvement** (In No Particular Order):

- Rational Knowledge (Intelligence): Collect and analyze data, build models, and conduct simulations to derive logical conclusions.
- Intuitive Knowledge (Imagination): Generate creative ideas through individual reflection, group discussions, and personal experience.
- Social Knowledge (Involvement): Gather insights into social factors, cultural norms, and stakeholder perspectives that influence decisions.

III. **Stage 3—Integration**:

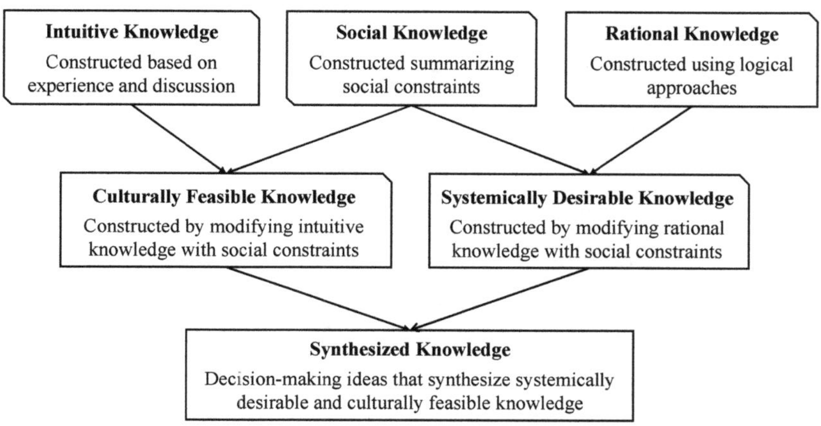

Fig. 5.12 The KTS framework (the knowledge construction diagram)

- Systemically Desirable Knowledge: Integrate rational knowledge with social constraints to create a balanced and optimal decision plan.
- Culturally Feasible Knowledge: Combine intuitive knowledge with social considerations to ensure that the plan is acceptable and practical for stakeholders.
- Synthesized Knowledge: Merge systemically desirable and culturally feasible knowledge to create a comprehensive decision-making proposal, ensuring a holistic, well-integrated final decision.

This framework provides a general approach to decision-making. The next section introduces a specialized framework for personal decisions, such as research and employment.

5.4.2 Strategy Story Creation

A strategy story is a narrative that communicates one's goals and intentions in a structured and persuasive manner. This section introduces a method for crafting a compelling strategy story using the KTS Framework.

Figure 5.13 illustrates the interconnected factors that influence personal decision-making. The diagram highlights the role of cognitive-mental factors in initiating the decision-making process, which follows this sequence: cognitive-mental → scientific-actual → social-relational.

Figure 5.14 presents the KTS Framework specifically adapted for strategy story creation. This framework offers a structured approach to integrating rational, intuitive, and social knowledge into a cohesive and convincing narrative.

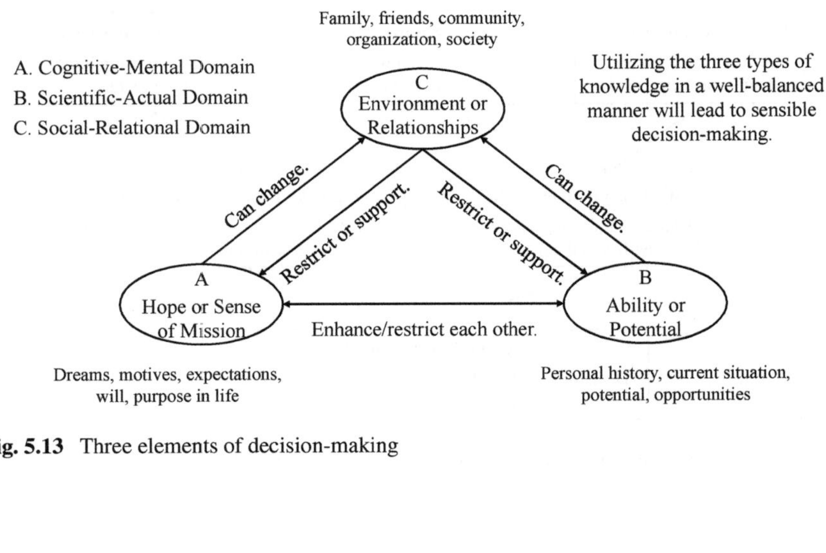

Fig. 5.13 Three elements of decision-making

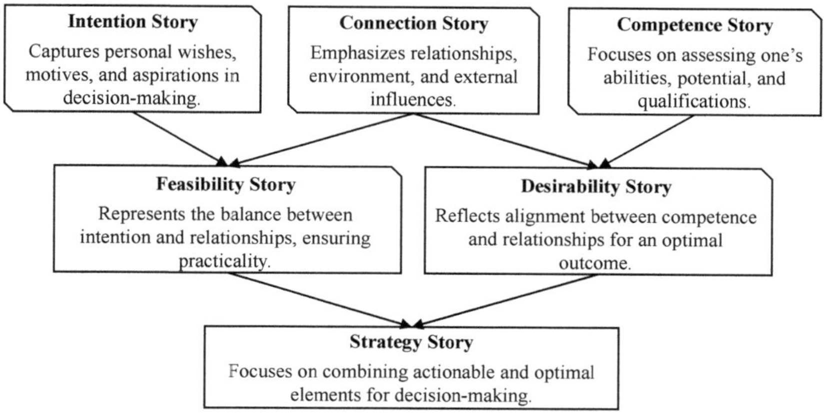

Fig. 5.14 KTS framework for strategy story creation

Example—Navigating a Parallel Career: Consider a middle-aged woman exploring the possibility of a parallel career. Popularized by management guru Peter Drucker, a parallel career involves pursuing a secondary profession alongside one's primary job. It goes beyond side hustles for extra income, encompassing activities that enrich one's life, such as volunteering or hobbies.

The increasing popularity of parallel careers can be attributed to several factors:

1. Economic Necessity: The need for additional income and skill development encourages individuals to explore side jobs or second careers.
2. Flexible Work Arrangements: The rise of freelancing and remote work enables people to pursue their passions while maintaining their primary employment.
3. Social Impact: Many individuals seek to contribute to society through meaningful work aligned with their interests.

In this scenario, the woman reflects: "*Now that my children are grown, I'm eager to find fulfilling work. Are there opportunities to contribute to society beyond my current job?*"

I. **Stage 1—Intervention**:
   - Define the Problem: Identify potential parallel career opportunities.
   - Develop a Knowledge Acquisition Plan:
     a. Aspirations: Define long-term career goals.
     b. Capability: Assess current skills and identify areas for improvement.
     c. Relationships: Explore personal and professional networks, including potential mentors and collaborators.

II. **Stage 2—Imagination, Intelligence, Involvement** (Think in This Order.):
   - **Intention Story**—*Leveraging Personal Passions and Values* (Imagination): My lifelong passion for intercultural exchange has allowed me to connect with people from diverse backgrounds. I have long aspired to use my

language skills professionally, such as in my company's trade department. While that opportunity has not materialized, the recent surge in foreign tourism in my city has inspired me to pursue a parallel career as a tour guide. I am confident in my ability to provide visitors with friendly, informative, and culturally sensitive guidance.

- **Competence Story**—*Maximizing Skills and Potential* (Intelligence): Having extensively studied English, I can communicate fluently and naturally. My deep knowledge of local history, culture, and attractions enables me to offer engaging tours. To further enhance my qualifications, I plan to pursue additional language training and tourism certifications. My ultimate goal is to become a fully licensed national guide-interpreter, complemented by a regional guide-interpreter certification.
- **Connection Story**—*Considering Social and Environmental Factors* (Involvement): With my children grown and my parents in good health, I have ample free time on weekends. My husband supports my desire to pursue work outside the home on weekends. My city, with its rich array of tourist attractions, has a growing demand for knowledgeable guides. While I cannot operate independently, I am eager to register with a volunteer organization, travel agency, guide company, or tourist association.

III. **Stage 3—Integration**:

- **Feasibility Story**—*A Culturally Feasible Plan* (Integration of Intention and Connection Stories): With my children now grown, I am ready to pursue my long-held passion for guiding foreign visitors. With my husband's support, I plan to integrate this role into my schedule. I consider this my calling because of my deep appreciation for language and culture. By respecting the customs of visitors from different countries, offering a warm smile, and providing clear, polite explanations, I hope to develop my skills and enrich my life.
- **Desirability Story**—*A Systemically Desirable Plan* (Integration of Competence and Connection Stories): With its many tourist attractions, my city has a rising demand for knowledgeable guides. I have a strong foundation in relevant languages and a deep understanding of local history and culture. To begin, I will register with a volunteer organization, a tour company, or a tourism association. Ultimately, I plan to become a nationally certified guide-interpreter by passing the Tourist English Proficiency Test and earning a regional guide-interpreter certification.
- **Strategy Story**—*A Synthesized Plan* (Integration of Feasibility and Desirability Stories): As tourism in my city grows, so does the demand for skilled guides catering to international visitors. With a passion for sharing my city's rich history and culture, I believe my language and communication skills make me an ideal candidate for this role. With my family's support, I am committed to refining my language skills and expanding my knowledge of local attractions. To gain hands-on experience, I will join a travel agency or

tourism association. I aim to pass the Tourist English Proficiency Test and become a certified local interpreter-guide.

Validity of the Story Synthesis: The synthesis of the above stories is validated as follows:

1. Feasibility Story—Integrating Intention and Connection Stories:
   - The Intention Story captures a passion for intercultural exchange, an aspiration to leverage language skills, and a desire to provide tourists with culturally sensitive and informative guidance.
   - The Connection Story addresses practical considerations, such as having free time, support from a spouse, and a growing demand for knowledgeable guides in the city.
   - The Feasibility Story merges these aspects, presenting a culturally feasible plan that balances personal motivations (from the Intention Story) with social and environmental factors (from the Connection Story). It builds on these foundations while incorporating a forward-looking perspective on skill development and life enrichment.
2. Desirability Story—Integrating Competence and Connection Stories:
   - The Competence Story highlights language proficiency, historical and cultural knowledge, and the ambition to earn certifications, such as national and regional guide-interpreter qualifications.
   - The Connection Story considers practical factors, including the increasing demand for guides, available free time on weekends, and plans to register with a volunteer or professional organization.
   - The Desirability Story combines these components, presenting a clear, actionable plan that integrates skills and aspirations (from the Competence Story) with social and environmental context (from the Connection Story). It outlines a step-by-step path toward achieving professional goals in tourism.
3. Strategy Story—Integrating Feasibility and Desirability Stories:
   - The Feasibility Story incorporates a passion for sharing culture and history, encouragement from family, and a desire to connect with diverse people while ensuring respectful and precise communication.
   - The Desirability Story includes an awareness of the increasing demand for guides, a commitment to skill enhancement, and a goal of obtaining certifications, such as passing the Tourist English Proficiency Test and becoming a certified local interpreter-guide.
   - The Strategy Story synthesizes these elements, forming a well-rounded plan that balances cultural enthusiasm and personal fulfillment (from the Feasibility Story) with professional aspirations, skill development, and certification goals (from the Desirability Story). This results in a comprehensive strategy for pursuing a parallel career as a tour guide.

Promotion Story: A Strategy Story outlines a systematic approach to achieving a goal. To bring this narrative to life, the woman crafts a Promotion Story (Nakamori et al., 2020), which focuses on expressing her intentions and goals while seeking cooperation from others. While the Strategy Story integrates various elements, the Promotion Story synthesizes them into a cohesive appeal for support.

Promotion Story*—Embracing a New Chapter with Community Support*:

- As tourism in our city grows, so does the demand for skilled guides who can accommodate international visitors. I am passionate about sharing our city's rich history and culture, and I believe my language and communication skills make me an ideal candidate for this role.
- With my family's unwavering support, I am committed to honing my language abilities and deepening my knowledge of local attractions. This journey is not just about personal growth; it is also about enriching tourists' experiences and contributing to our community's vibrant cultural exchange.
- To gain practical experience, I will register with a travel agency or tourism association. My goal is to pass the Tourist English Proficiency Test and become a certified local interpreter-guide. Yet, I cannot achieve this alone.
- I kindly ask for your support and encouragement as I embark on this exciting new chapter. Your guidance, advice, and connections will be invaluable in helping me make a meaningful impact on our community. Together, we can create memorable experiences for visitors and showcase our city's beauty to the world.
- Thank you for believing in me and for being part of this journey.

5.5 End-of-Chapter Problems

5.5.1 Problem 5.1: Synthetic Thinking

Synthetic thinking synthesizes diverse knowledge sources—such as data, trends, and insights—to generate novel ideas and a comprehensive understanding beyond conventional problem-solving.

Understanding synthetic thinking requires exploring related conceptual frameworks. One relevant framework is *design thinking*, particularly its ideation stage, which consists of two complementary processes:

- *Divergent Thinking*: Encourages brainstorming and open-ended ideation to generate a wide range of possible solutions.
- *Convergent Thinking*: Analyzes and evaluates ideas, narrowing them down to the most feasible and promising solutions.

Table 5.3 highlights the key differences between these approaches.

Task: Compare the differences between divergent-convergent thinking and synthetic thinking by completing the following steps:

Table 5.3 Key differences between divergent-convergent thinking and synthetic thinking

Aspect	Divergent-convergent thinking	Synthetic thinking
Focus	A process for solving a specific problem or challenge	A method for creating new and innovative ideas
Approach	A cyclical process of idea generation and evaluation	An approach that integrates knowledge from various domains
Emphasis	A strategy for selecting the best solution from existing options	A technique for generating novel insights and unique solutions
Scope	A framework primarily focused on problem-solving	A broader approach that extends beyond problem-solving to foster new knowledge
Nature	A structured, linear process that follows the sequence: diverge → converge	A holistic and interconnected approach

1. Analyze how these approaches differ in terms of focus and process.
2. Illustrate these differences using an example: planning a summer menu for a sushi restaurant.

5.5.2 Problem 5.2: AlphaGo

AlphaGo, developed by Google DeepMind, is a groundbreaking AI program that has transformed the landscape of Go. By utilizing deep learning and reinforcement learning, AlphaGo attained a superhuman level of play, defeating some of the world's top professional players.

AlphaGo has several key features, including:

- Deep Neural Networks: AlphaGo leveraged deep neural networks to analyze game positions, recognize patterns, and predict optimal moves.
- Reinforcement Learning: Through self-play, AlphaGo played millions of games against itself, continuously refining its strategies and improving its decision-making capabilities.

The significance of AlphaGo lies in its ability to:

- Overcoming a Grand Challenge: The complexity of Go, with its astronomical number of possible moves (exceeding the number of atoms in the universe), posed a significant challenge for AI. AlphaGo's success demonstrated that AI could excel in domains traditionally associated with human intuition, creativity, and strategic reasoning.
- Advancing AI Capabilities: AlphaGo's achievements have pushed the boundaries of artificial intelligence, showcasing the power of deep learning and reinforcement learning in tackling complex problems.

AlphaGo's victories over human champions marked a historic milestone in AI development, highlighting the extraordinary potential of machine learning in mastering a game of profound strategic depth and complexity.

Task: What are the fundamental differences between human professional Go players and AlphaGo?

5.5.3 Problem 5.3: Tacit Integration

Michael Polanyi, a philosopher of science, introduced the concept of tacit integration in *The Tacit Dimension.* He argued that humans possess an intuitive and deeply embedded understanding of things that often eludes conscious articulation—such as recognizing a friend's face.

AI-based facial recognition differs significantly from human recognition. AI follows a structured, algorithmic process:

1. Face Detection: The system identifies a face within an image or video by analyzing patterns such as the positioning of the eyes, nose, and mouth.
2. Feature Extraction: It extracts key facial features (e.g., eye distance, nose shape, and facial contours) to create a unique digital representation, or faceprint.
3. Face Matching: The faceprint is compared to a database of known faces using similarity algorithms.
4. Verification or Identification: The system either verifies an identity by matching the faceprint to a stored template or identifies an unknown individual by searching the database.

Facial recognition in AI is a complex, multi-step process that relies on explicit feature analysis rather than intuitive perception.

Task: Polanyi stated that humans recognize their friends' faces through *tacit integration.* How does human facial recognition differ from AI-based recognition?

5.5.4 Problem 5.4: Knowledge Triad Synthesis

Suppose you are developing a summer menu for a sushi restaurant. You have gathered the following insights:

1. Objective Facts (Rational Ideas)

    - Certain types of sushi sell less in summer than in winter, but this is not due to a decline in customer demand. Instead, it represents an opportunity loss because seasonal winter fish are underutilized in summer.
    - Importing and serving winter fish in summer is necessary, even if it incurs slightly higher costs.

- Regarding the side menu, warm seafood soups and noodles are in high demand during winter, whereas vegetable salads and tempura are more popular in summer.

2. Wishful Thinking (Intuitive Ideas)

- Creative menu ideas can emerge by combining various concepts, such as children on summer vacation, fresh summer fish, high-end winter fish, popular dishes, and profitability.
- During summer vacation, children's sushi sets could be introduced, featuring popular items like omelet (egg sushi) and chicken.
- To maintain the availability of popular fish year-round, options include importing or using frozen fish, even if it results in lower profit margins.
- Buying fresh fish in bulk during peak seasons can help improve profitability.

3. Customer Expectations (Social Ideas)

- Customers prioritize fresh, seasonal, and high-quality fish.
- While there is strong demand for high-quality winter fish, customers expect it to be reasonably priced.
- There is growing interest in a more varied side menu, including seafood soups, summer vegetable salads, and seasonal tempura.
- Affordable, kid-friendly sushi options are in demand, particularly during summer vacations when families dine out more often.

Task: Synthesize these ideas using the Knowledge Triad Synthesis Framework.

1. Derive *systemically desirable ideas* by reinforcing or restricting rational ideas with social ideas.
2. Derive *culturally feasible ideas* by reinforcing or restricting intuitive ideas with social ideas.
3. Synthesize *systemically desirable ideas* and *culturally feasible ideas* to create a balanced and actionable summer menu plan.

5.5.5 Problem 5.5: Strategy Story

A strategy story is a persuasive narrative that outlines an action plan for achieving a desired outcome, emphasizing its benefits, unique value, and key selling points.

Task: Develop a strategy story for your research, career choice, or other topic.

1. Define Your Purpose: Create Intention, Competence, and Connection Stories to establish motivation, expertise, and external influences.
2. Construct Your Strategy Story: Apply the Knowledge Triad Synthesis Framework to integrate intuitive ideas, rational insights, and social considerations into a compelling and cohesive narrative.

References

Brown, T. (2008). Design thinking. *Harvard Business Review, 86*(6), 84–92.

Checkland, P. B. (1981). *Systems thinking, systems practice.* Wiley.

Csikszentmihalyi, M. (2015). *The systems model of creativity: The collected works of Mihaly Csikszentmihalyi.* Springer.

Drucker, P. F. (1954). *The practice of management.* HarperCollins.

Drucker, P. F. (1973). *Management: Tasks, responsibilities.* HarperCollins.

Drucker, P. F. (1985). *Innovation and entrepreneurship.* HarperCollins.

Forrester, J. W. (1961). *Industrial dynamics.* MIT Press.

Forrester, J. W. (1969). *Urban dynamics.* Pegasus Communications.

Forrester, J. W. (1971). *World dynamics.* Wright-Allen Press.

Gu, J. F., & Tang, X. J. (2000). Designing a water resources management decision support system: An application of the WSR approach. *Systemic Practice and Action Research, 13*, 70–77.

Gu, J. F., & Tang, X. J. (2004). Wu-li Shi-li Ren-li system approach to a major project on the research of meta-synthesis system approach. *International Journal of Knowledge and Systems Sciences, 1*(1), 70–77.

Gu, J. F., & Zhu, Z. C. (1995). The Wuli-Shili-Renli approach: An oriental systems methodology. In G. Midgley & J. Wiley (Eds.), *Systems methodology I: Possibilities for cross-cultural learning and integration* (pp. 29–38). University of Hull.

Gu, J. F., & Zhu, Z. C. (2000). Knowing Wuli, sensing Shili, caring Renli: Methodology of the WSR approach. *Systems Practice and Action Research, 13*(1), 11–20.

Nakamori, Y. (2019). *Knowledge construction methodology: Fusing systems thinking and knowledge management.* Springer.

Nakamori, Y. (2021). *Knowledge technology: Converting data and information into valuable ideas.* Springer.

Nakamori, Y., Udomvitid, K., Chanyachatchawan, S., & Karnjana, J. (2020). Creating promotion stories with knowledge management reinforced by systems thinking. *Journal of Systems Science and Systems Engineering, 29*(2), 163–183.

Polanyi, M. (1958). *Personal knowledge: Towards a post-critical philosophy.* University of Chicago Press.

Polanyi, M. (1966). *The tacit dimension.* Routledge & Kegan Paul.

Qian, X. S., Yu, J. Y., & Dai, R. W. (1990). A new discipline of science—The study of open complex giant system and its methodology. *Chinese Journal of Systems Engineering and Electronics, 4*(2), 2–12.

Schumpeter, J. A. (1912). *Theorie der wirtschaftlichen Entwicklung. Translation*: Opie, R. (1934). *The theory of economic development: An inquiry into profits, capital, credit, interest, and the business cycle.* Translation Publishers.

Schumpeter, J. A. (1942). *Capitalism, socialism and democracy.* Harper & Brothers.

Wierzbicki, A. P., & Nakamori, Y. (Eds.). (2007). *Creative environments: Issues of creativity support for the knowledge civilization Aga.* Springer.

Zhu, Z. C. (2023). WSR: A Chinese systems approach. *Journal of Systems Thinking, 3*, 1–13.

Further Reading on Synthetic Thinking

de Bono, E. (2015). *Lateral thinking: Creativity step by step.* Harper Colophon. This book introduces lateral thinking, a creative problem-solving approach that complements traditional logical reasoning. It provides practical tools to think outside the box.

Dennett, D. C. (2014). *Intuition pumps and other tools for thinking*. Penguin. A collection of cognitive tools to tackle complex topics such as evolution, consciousness, and free will, offering practical strategies for clearer thinking.

Kahneman, D. (2011). *Thinking, fast and slow*. Penguin. This work explores the dual systems of thought: the fast, intuitive system and the slow, logical system, offering insights into how these systems shape decision-making.

Martin, R. L. (2007). *The opposable mind: How successful leaders win through integrative thinking*. Harvard Business Review Press. This book presents the concept of integrative thinking, explaining how successful leaders combine opposing approaches to generate innovative solutions.

Parrish, S. (2018). *The great mental models: General thinking concepts*. Latticework Publishing. A guide to essential mental models that improve decision-making and problem-solving by offering a structured framework for understanding complex ideas.

Williams, A. J. (2014). *Critical thinking: Decision making with smarter logic and intuition*. This book explains critical thinking in depth, demonstrating how to achieve better decisions by balancing logic with intuitive insights.

Chapter 6
Example Answers to End-of-Chapter Problems

Abstract This chapter provides example answers to the end-of-chapter problems from Chaps. 1 to 5. These problems are designed to deepen the reader's understanding of the concepts covered in each chapter and are presented in a way that complements the chapter material. Except for specific mathematical problems, the answers are not unique and should be used solely as a reference. Nonetheless, these example answers aim to enhance the reader's comprehension of each chapter's content.

6.1 Example Answers to Problems in Chapter 1

6.1.1 Example Answer to Problem 1.1: Expected Value Paradox

1. Lottery

Several factors explain why people buy lottery tickets despite the low probability of winning:

- *The Dream of Winning Big*: The appeal lies in the possibility of winning a life-changing sum of money, and imagining "What if I win?" evokes joy, hope, and a sense of possibility. This dream-like attraction motivates people to buy tickets, even with extremely low odds.
- *Entertainment Value*: Lottery tickets provide excitement and thrill, making them a form of entertainment. The anticipation while waiting for the results and imagining how life might change creates moments of joy. They also foster social interaction, such as discussing plans with friends or family about what to do if they win.
- *Psychological Effects*: Buying a lottery ticket triggers the release of dopamine, a neurotransmitter associated with happiness and pleasure. As a result, purchasing a lottery ticket becomes a positive experience, fostering optimism and a temporary sense of control over one's fate.

Y. Nakamori, *Sensible Decision-Making*, https://doi.org/10.1007/978-981-96-6811-3_6

- *Cognitive Biases*: Psychological biases, such as the anchoring bias (the tendency to rely too heavily on the first piece of information encountered), lead people to overestimate the value of lottery tickets. For instance, prominently displayed jackpot amounts influence perceptions, causing individuals to disregard the low probability of winning.

In summary, purchasing lottery tickets is driven by psychological, emotional, and social factors that extend beyond purely economic rationality.

2. St. Petersburg Paradox

Despite the infinite expected value of the St. Petersburg Paradox, most people are unwilling to pay a considerable amount to play. Several factors explain this apparent contradiction:

- *Diminishing Marginal Utility*: As wealth increases, the additional satisfaction (utility) gained from each extra dollar diminishes. A billion-dollar gain might bring happiness, but a trillion-dollar gain adds far less incremental joy. Bernoulli's concept of expected utility—rather than expected value—helps explain why people are reluctant to pay high amounts for the game.
- *Risk Aversio*n: Most people prefer certainty over risk, especially when facing the possibility of significant losses. While the game's payouts are theoretically massive, the extremely low probability of achieving these rewards makes the risk unappealing in practice.
- *Cognitive Limitations*: Humans struggle to intuitively grasp extreme probabilities and large payoffs. Our decision-making processes cannot effectively handle scenarios involving vast uncertainty scales, leading to hesitation in placing high bets.

Conclusion: Expected value alone is often insufficient for decision-making in uncertain real-world situations. Psychological factors such as risk aversion, diminishing marginal utility, and cognitive limitations significantly influence how people evaluate choices. These considerations explain why people deviate from expected value norms, particularly in scenarios like lotteries or the St. Petersburg Paradox.

6.1.2 Example Answer to Problem 1.2: Hurwicz Criterion

Table 6.1 presents a revised version of Table 1.6 to analyze investment decisions for different values of α. The Weighted sum column in Table 6.1 shows the Hurwicz criteria $S_i (i = 1,2, \cdots, 11)$ for 11 different diversification strategies. The last column displays the difference between S_{11} and the other criteria.

From Table 6.1, we observe the following:

1. When $0 \leq \alpha < 0.5$, $S_k < 10 = S_5$ for all $k \neq 5$

   → The optimal investment is \$400 in the concert and \$600 in the movie.

Table 6.1 Investment and weighted sum (Unit: $10)

Investment		Profit		Weighted sum	$S_{11} - S_k$
Concert	Movie	Max	Min	$\alpha \times \text{Max} + (1-\alpha) \times \text{Min}$	
0	100	70	−50	$S_1 = 120\alpha - 50$	$60\alpha - 30$
10	90	55	−35	$S_2 = 90\alpha - 35$	$90\alpha - 45$
20	80	40	−20	$S_3 = 60\alpha - 20$	$120\alpha - 60$
30	70	25	−5	$S_4 = 30\alpha - 5$	$150\alpha - 75$
40	60	**10**	**10**	$S_5 = 10$	$180\alpha - 90$
50	50	25	−5	$S_6 = 30\alpha - 5$	$150\alpha - 75$
60	40	40	−20	$S_7 = 60\alpha - 20$	$120\alpha - 60$
70	30	55	−35	$S_8 = 90\alpha - 35$	$90\alpha - 45$
80	20	70	−50	$S_9 = 120\alpha - 50$	$60\alpha - 30$
90	10	85	−65	$S_{10} = 150\alpha - 65$	$30\alpha - 15$
100	0	**100**	**−80**	$S_{11} = 180\alpha - 80$	0

2. When $\alpha = 0.5,\ S_k = 10$ for all k

   → Any combination of investments is equally optimal.

3. When $0.5 < \alpha \leq 1,\ S_k < S_{11}$ for all $k \neq 11$

   → The optimal investment is $1000 in the concert.

Conclusion: Based on the value of α, the following investments are recommended:

- For $\alpha < 0.5$: The Hurwicz criterion favors a pessimistic approach aligned with the Maximin strategy, recommending diversification (investment in both the concert and the movie).
- For $\alpha > 0.5$: The criterion shifts toward an optimistic approach similar to the Maximax strategy, prioritizing maximum potential gain (concert-only investment).
- For $\alpha = 0.5$: Any investment combination is optimal since it balances optimism and pessimism equally.

This problem demonstrates how the Hurwicz criterion integrates optimism and pessimism in decision-making under uncertainty, with the value of α determining the balance between the two.

6.1.3 Example Answer to Problem 1.3: Minimax Regret

Table 6.2 outlines a regret matrix, derived from Table 1.6, to identify the investment option with the lowest maximum regret. This matrix shows the regret values for two scenarios—Sunny/Cloudy and Rainy/Snowy—as well as the maximum regret associated with each investment choice.

Steps to apply the Minimax Regret criterion are as follows:

1. Calculate Regret: For each investment level, determine the difference between the outcome and the best possible outcome for each weather scenario.
2. Determine Maximum Regret: Identify the largest regret for each investment across all scenarios (as listed in the "Max Regret" column).
3. Select the Optimal Investment: Choose the investment option with the smallest value in the "Max Regret" column.

From Table 6.2:

- The smallest maximum regret is 75, corresponding to an investment of \$500 in the concert and \$500 in the movie.

Conclusion: Using the Minimax Regret criterion, the optimal investment is:

- \$500 in the concert and \$500 in the movie.
- If the weather is Sunny/Cloudy, this choice yields a profit of \$25.
- If the weather is Rainy/Snowy, this choice minimizes the loss to \$5.

This strategy ensures the lowest maximum regret across all potential weather scenarios.

Table 6.2 Diversified investment and regrets (Unit: \$10)

Investment		Profit matrix		Regret matrix		Max regret
Concert	Movie	Sunny/cloudy	Rainy/snowy	Sunny/cloudy	Rainy/snowy	
0	100	−50	70	150	0	150
10	90	−35	55	135	15	135
20	80	−20	40	120	30	120
30	70	−5	25	105	45	105
40	60	10	10	90	60	90
50	**50**	**25**	**−5**	**75**	**75**	**75**
60	40	40	−20	60	90	90
70	30	55	−35	45	105	105
80	20	70	−50	30	120	120
90	10	85	−65	15	135	135
100	0	100	−80	0	150	150

6.1.4 Example Answer to Problem 1.4: Decision-Making Criteria

1. Expected Value Maximization

Use this criterion when:

- The probabilities of different outcomes are known.
- Outcomes can be quantified in numerical terms.

This approach is ideal for repeated decisions where the goal is to maximize the long-term average payoff or utility (e.g., financial investments or gambling scenarios).

2. Maximizing the Worst-Case Outcome (Maximin):

Use this criterion when:

- You are risk-averse and prioritize minimizing losses.
- There is significant uncertainty, and protection against the worst-case scenario is a priority.

Maximin is a conservative strategy, suitable for highly uncertain environments or situations where failure has severe consequences (e.g., disaster response planning).

3. Maximizing the Best Possible Outcome (Maximax)

Use this criterion when:

- You are optimistic and willing to take risks for the highest reward.
- There is a high potential for substantial gains, and you can tolerate the risk of significant losses.

Maximax is an aggressive strategy, often applied to situations with large potential payoffs (e.g., entrepreneurial ventures or groundbreaking innovations).

4. Minimizing Regret (Minimax Regret)

Use this criterion when:

- You aim to avoid future regret from making a suboptimal decision.
- You are concerned about the difference between your outcome and the best possible outcome.

Minimax Regret is particularly effective in highly uncertain scenarios where mitigating potential disappointment is essential (e.g., career or life-altering decisions).

6.1.5 *Example Answer to Problem 1.5: Prisoner's Dilemma*

Situations resembling the Prisoner's Dilemma arise when there is a conflict between individual and collective interests. Two main factors contribute to this phenomenon:

- Individual Rationality: Players focus on maximizing their short-term gains, often at the expense of long-term benefits.
- Incomplete Information: Players frequently lack complete knowledge of their opponent's actions or intentions, creating uncertainty.

Even when cooperation would lead to better outcomes for all parties, the fear of defection often drives individuals to prioritize the immediate rewards of acting in their self-interest.

The Prisoner's Dilemma can be observed in real-world scenarios, such as:

- Environmental issues: Companies may achieve short-term financial gains by neglecting environmental protection, but the resulting ecological damage affects everyone—including those very companies—over time.
- Arms race: When one country increases its military capabilities, others do the same, leading to escalating military expenditures without enhancing overall security.

6.2 Example Answers to Problems in Chapter 2

6.2.1 *Example Answer to Problem 2.1: Sensibility in Decision-Making*

Sensibility plays a crucial role in decision-making. It allows us to consider nuanced factors beyond pure logic and data, providing intuitive insights that complement rational analysis. By integrating sensibility-based evaluations, we can recognize patterns, anticipate risks, and make informed judgments based on emotions and past experiences. Combining logical and intuitive approaches enables us to make more well-rounded and effective choices.

1. Comparing Rational and Sensibility-Oriented Approaches in Lunch Decisions:
    - *Rational approach*: A rational decision-making process for choosing lunch might involve the following considerations:
        - Nutritional balance: Analyzing the macronutrient and micronutrient composition of different meal options to ensure they meet dietary needs.
        - Energy efficiency: Selecting a meal that provides sufficient energy for the rest of the day's activities while minimizing potential energy crashes.
        - Cost-effectiveness: Comparing prices of available options to ensure the best nutritional value for the cost.

  - Health impact: Avoiding foods high in unhealthy fats, sugars, or additives that could lead to long-term health concerns.

- *Sensibility-oriented approach*: In addition to the above logical considerations, sensibility introduces factors such as:
  - Practicality: How much time is available to prepare or purchase lunch? If a healthy option is time-consuming, is there a reasonable alternative?
  - Emotional state: Am I craving something comforting, refreshing, or indulgent? Will this meal improve my mood or provide psychological satisfaction?
  - Long-term impact: Will this choice make me feel physically and mentally well throughout the day? Will I regret not choosing a more enjoyable meal?

By incorporating these factors, one can make a decision that is not only nutritionally sound and cost-effective but also satisfying and aligned with current needs.

2. Other Examples of Sensibility in Action:
   - Choosing clothes:
     - *Rational approach*: Assessing the clothing's durability, material quality, and cost-effectiveness to maximize long-term value.
     - *Sensibility-oriented approach*: Considering comfort, personal style, and how the outfit aligns with the day's activities and weather conditions.
   - Investing money:
     - *Rational approach*: Conducting financial analyses, including expected returns, market trends, and risk assessments, to determine the most profitable investment.
     - *Sensibility-oriented approach*: Factoring in personal risk tolerance, lifestyle stability, and long-term financial goals to ensure peace of mind and financial security.
   - Accepting a job offer:
     - *Rational approach*: Comparing salaries, benefits, and career advancement opportunities in a purely analytical manner.
     - *Sensibility-oriented approach*: Evaluating work-life balance, company culture, commute, and overall job satisfaction beyond just financial compensation.

Conclusion: Sensibility is an essential complement to rationality in decision-making. It ensures that practical considerations, emotions, and long-term effects are taken into account alongside logical reasoning. By integrating both aspects, we can make more balanced, realistic, and sustainable choices in everyday life.

6.2.2 Example Answer to Problem 2.2: Affective Engineering

Affective Engineering and Emotional Design focus on the five human senses and emotions, exploring how they influence product and service evaluations and final decision-making. These fields enhance decision-making by incorporating an intuitive, human-centered perspective into traditional logical approaches.

1. *Contributions to Decision-Making*:
   - Optimizing Product Development: By analyzing consumer responses to sensory and emotional elements—such as shape, color, texture, and scent—products can be refined to maximize their appeal. Integrating emotional elements into design helps establish a brand's identity and image, creating a memorable brand experience that resonates with consumers.
   - Enhancing Marketing Strategies: Emotional appeals in advertisements and promotions can significantly boost engagement and effectiveness. Store designs and product displays that stimulate the senses can heighten consumer interest, increase purchase intent, and even trigger impulse buys.
   - Improving Service Design: Enhancing service interactions, ambiance, and spatial design can increase customer satisfaction and encourage repeat visits. Analyzing the emotional aspects of customer experiences helps identify areas for improvement and enables businesses to create more personalized and engaging interactions.
2. *Roles in Decision-Making*:
   - Integrating Quantitative and Qualitative Data: Sensibility-based evaluations often combine quantitative data (e.g., physiological response measurements, surveys) with qualitative data (e.g., interviews, observations). This integration provides deeper insights into consumer preferences and behaviors, leading to more informed decision-making.
   - Justifying Intuitive Judgments: Many decisions in product and service design rely on intuition. Affective Engineering provides a scientific framework to validate these intuitive judgments, making them more reliable and justifiable in business and design contexts.
   - Creating New Value: Affective Engineering helps uncover latent consumer needs and desires, driving the development of innovative products and services. By incorporating emotional and sensory dimensions into existing offerings, companies can enhance their perceived value and strengthen brand loyalty.

Conclusion: Affective Engineering and Emotional Design bridge the gap between logic and emotion in decision-making. They enable businesses to address both functional and emotional needs, leading to more innovative, impactful, and consumer-centric outcomes in product development, marketing, and service design.

6.2.3 *Example Answer to Problem 2.3: Possibility Models*

Suppose discrete possibility models are constructed as shown in Tables 6.3, 6.4, 6.5, 6.6, and 6.7. Please note that these models are highly subjective. The bold numbers in these tables represent the customer's desired levels:

- "Soft": Level 2 ($l = 2$)
- "Very warm": Level 7 ($l = 7$)
- "A little for elderly": Level 5 ($l = 5$)
- "For women": Level 6 ($l = 6$)

Table 6.3 Discrete possibility models for Cup o_1

Axis	Left word	$l = 1$	$l = 2$	$l = 3$	$l = 4$	$l = 5$	$l = 6$	$l = 7$	Right word
w_1	Soft	0.6	**0.8**	1.0	0.8	0.6	0.4	0.2	Hard
w_2	Cool	0.2	0.4	0.6	0.8	1.0	0.8	**0.6**	Warm
w_3	For youth	0.8	1.0	0.8	0.6	**0.4**	0.2	0.0	For elderly
w_4	For men	0.0	0.2	0.4	0.6	0.8	**1.0**	0.8	For women

Table 6.4 Discrete possibility models for Cup o_2

Axis	Left word	$l = 1$	$l = 2$	$l = 3$	$l = 4$	$l = 5$	$l = 6$	$l = 7$	Right word
w_1	Soft	0.6	**0.8**	1.0	0.8	0.6	0.4	0.2	Hard
w_2	Cool	0.0	0.2	0.4	0.6	0.8	1.0	**0.8**	Warm
w_3	For youth	0.4	0.6	0.8	1.0	**0.8**	0.6	0.4	For elderly
w_4	For men	0.0	0.0	0.2	0.4	0.6	**0.8**	1.0	For women

Table 6.5 Discrete possibility models for Cup o_3

Axis	Left word	$l = 1$	$l = 2$	$l = 3$	$l = 4$	$l = 5$	$l = 6$	$l = 7$	Right word
w_1	Soft	0.8	**1.0**	0.8	0.6	0.4	0.2	0.0	Hard
w_2	Cool	0.0	0.2	0.4	0.6	0.8	1.0	**0.8**	Warm
w_3	For youth	0.6	0.8	1.0	0.8	**0.6**	0.4	0.2	For elderly
w_4	For men	0.2	0.4	0.6	0.8	1.0	**0.8**	0.6	For women

Table 6.6 Discrete possibility models for Cup o_4

Axis	Left word	$l = 1$	$l = 2$	$l = 3$	$l = 4$	$l = 5$	$l = 6$	$l = 7$	Right word
w_1	Soft	0.2	**0.4**	0.6	0.8	1.0	0.8	0.6	Hard
w_2	Cool	0.2	0.4	0.6	0.8	1.0	0.8	**0.6**	Warm
w_3	For youth	0.0	0.2	0.4	0.6	**0.8**	1.0	0.8	For elderly
w_4	For men	0.2	0.4	0.6	0.8	1.0	**0.8**	0.6	For women

Table 6.7 Discrete possibility models for Cup o_5

Axis	Left word	$l=1$	$l=2$	$l=3$	$l=4$	$l=5$	$l=6$	$l=7$	Right word
w_1	Soft	0.6	**0.8**	1.0	0.8	0.6	0.4	0.2	Hard
w_2	Cool	0.0	0.0	0.2	0.4	0.6	0.8	**1.0**	Warm
w_3	For youth	0.0	0.2	0.4	0.6	**0.8**	1.0	0.8	For elderly
w_4	For men	0.4	0.6	0.8	1.0	0.8	**0.6**	0.4	For women

Table 6.8 Compatibility table with customer requirements

Object	Soft ($l=2$)	*Very* warm ($l=7$)	*A little* for elderly ($l=5$)	For women ($l=6$)
Cup o_1	0.8	0.6	0.4	1.0
Cup o_2	0.8	0.8	0.8	0.8
Cup o_3	1.0	0.8	0.6	0.8
Cup o_4	0.4	0.6	0.8	0.8
Cup o_5	0.8	1.0	0.8	0.6

Table 6.8 presents the compatibility table based on Tables 6.3, 6.4, 6.5, 6.6, and 6.7.

6.2.4 *Example Answer to Problem 2.4: OWA Operators*

Step 1. Transcribe the Compatibility Table: The compatibility table (Table 6.8) is rewritten as Table 6.9, incorporating the necessary symbols.

Step 2. Determine Weights Using Yager's Method: Using the membership functions in Table 2.7, we calculate the weights when the number of attributes is four and obtain Table 6.10.

Step 3. Calculate the Overall Evaluation Values: The rating values are sorted in descending order. If there is a tie, the option with the lower subscript is chosen. The overall evaluation score for each cup under each condition is calculated as follows:

Table 6.9 The compatibility table with necessary symbols

Object	w_1 : Soft	w_2 : Very warm	w_3 : A little for elderly	w_4 : For women
Cup o_1	$s_{11}=0.8$	$s_{12}=0.6$	$s_{13}=0.4$	$s_{14}=1.0$
Cup o_2	$s_{21}=0.8$	$s_{22}=0.8$	$s_{23}=0.8$	$s_{24}=0.8$
Cup o_3	$s_{31}=1.0$	$s_{32}=0.8$	$s_{33}=0.6$	$s_{34}=0.8$
Cup o_4	$s_{41}=0.4$	$s_{42}=0.6$	$s_{43}=0.8$	$s_{44}=0.8$
Cup o_5	$s_{51}=0.8$	$s_{52}=1.0$	$s_{53}=0.8$	$s_{54}=0.6$

Table 6.10 Weights of aggregating partial ratings

Request	Weight
A. Meets all the requests	$\mu_1 = 0, \mu_2 = 0, \mu_3 = 0, \mu_4 = 1$
B. Meets as many requests as possible	$\mu_1 = 0, \mu_2 = 0, \mu_3 = 1/2, \mu_4 = 1/2$
C. Meets at least half of the requests	$\mu_1 = 1/2, \mu_2 = 1/2, \mu_3 = 0, \mu_4 = 0$
D. Meets at least one request	$\mu_1 = 1, \mu_2 = 0, \mu_3 = 0, \mu_4 = 0$

A. Meets All the Requests

- $E_1 = \mu_4 \times s_{13} = 0.4$
- $E_2 = \mu_4 \times s_{21} = 0.8$
- $E_3 = \mu_4 \times s_{33} = 0.6$
- $E_4 = \mu_4 \times s_{41} = 0.4$
- $E_5 = \mu_4 \times s_{54} = 0.6$

B. Meets as Many Requests as Possible

- $E_1 = \mu_3 \times s_{12} + \mu_4 \times s_{13} = 0.5$
- $E_2 = \mu_3 \times s_{22} + \mu_4 \times s_{21} = 0.8$
- $E_3 = \mu_3 \times s_{32} + \mu_4 \times s_{33} = 0.7$
- $E_4 = \mu_3 \times s_{42} + \mu_4 \times s_{41} = 0.5$
- $E_5 = \mu_3 \times s_{51} + \mu_4 \times s_{54} = 0.7$

C. Meets at Least Half of the Requests

- $E_1 = \mu_1 \times s_{14} + \mu_2 \times s_{11} = 0.9$
- $E_2 = \mu_1 \times s_{21} + \mu_2 \times s_{22} = 0.8$
- $E_3 = \mu_1 \times s_{31} + \mu_2 \times s_{32} = 0.9$
- $E_4 = \mu_1 \times s_{43} + \mu_2 \times s_{44} = 0.8$
- $E_5 = \mu_1 \times s_{52} + \mu_2 \times s_{51} = 0.9$

D. Meets at Least One Request

- $E_1 = \mu_1 \times s_{14} = 1.0$
- $E_2 = \mu_1 \times s_{21} = 0.8$
- $E_3 = \mu_1 \times s_{31} = 1.0$
- $E_4 = \mu_1 \times s_{43} = 0.8$
- $E_5 = \mu_1 \times s_{52} = 1.0$

Summary of Results:

- Cup o_2 is the best choice when prioritizing complete fulfillment of all requests since it consistently meets every criterion at a high level.
- Cup o_2 is also the best option for satisfying as many requests as possible because it maintains a strong balance across multiple factors.
- Cups o_1, o_3, and o_5 are suitable for meeting at least half of the requests as they score highly on some key criteria, though not all.

- Cups o_1, o_3, and o_5 are also viable options if satisfying at least one request is sufficient, as each fully aligns with at least one of the customer's preferences.

In summary, Cup o_2 is the most well-rounded recommendation, while Cups o_1, o_3, and o_5 serve as acceptable alternatives depending on the required level of compatibility.

6.2.5 Example Answer to Problem 2.5: Choquet Integral

We begin by defining the weights of the four criteria:

$$w_1 : \text{Soft}, \;\; w_2 : \text{Very warm}, \;\; w_3 : \text{A little for elderly}, \;\; w_4 : \text{For women}$$

Then, we apply the Choquet integral to compute the overall evaluation values for the five cups and determine the best recommendation.

Step 1: Define Weights.

- Weights for Single Attribute Evaluation

$$\mu(\{w_1\}) = 0.4, \;\; \mu(\{w_2\}) = 0.2, \;\; \mu(\{w_3\}) = 0.2, \;\; \mu(\{w_4\}) = 0.2$$

- Weights for Pairs of Attributes

$$\mu(\{w_1, w_2\}) = 0.4, \;\; \mu(\{w_1, w_3\}) = 0.4, \;\; \mu(\{w_1, w_4\}) = 0.4$$

$$\mu(\{w_2, w_3\}) = 0.2, \;\; \mu(\{w_2, w_4\}) = 0.2, \;\; \mu(\{w_3, w_4\}) = 0.7$$

- Weights for Triplets of Attributes

$$\mu(\{w_1, w_2, w_3\}) = 0.4, \;\; \mu(\{w_1, w_2, w_4\}) = 0.4$$

$$\mu(\{w_1, w_3, w_4\}) = 1.0, \;\; \mu(\{w_2, w_3, w_4\}) = 0.7$$

- Weights for Empty and Full Sets:

$$\mu(\varnothing) = 0.0, \;\; \mu(\{w_1, w_2, w_3, w_4\}) = 1.0$$

The justification for the weights is as follows:

- Condition 1: $\mu(\{w_1\}) = 0.4$ reflects the importance of "Soft" as the primary criterion.

- Condition 2: $\mu(\{w_3, w_4\}) = 0.7$ assigns a high weight when both "A little for elderly" and "For women" are satisfied.
- Condition 3: $\mu(\{w_1, w_3, w_4\}) = 1.0$ ensures the highest weight for fulfilling "Soft," "A little for elderly," and "For women."
- Other weights are set to the smallest values that satisfy monotonicity.

Step 2: Compute the Choquet integrals.

Using the compatibility scores s_{ij}, the Choquet integral E_m for each cup o_m is calculated as follows:

- $$\begin{aligned} E_1 &= \mu(\{w_4\}) \times (s_{14} - s_{11}) + \mu(\{w_1, w_4\}) \times (s_{11} - s_{12}) \\ &\quad + \mu(\{w_1, w_2, w_4\}) \times (s_{12} - s_{13}) + \mu(\{w_1, w_2, w_3, w_4\}) \times s_{13} \\ &= 0.2 \times 0.2 + 0.4 \times 0.2 + 0.4 \times 0.2 + 1.0 \times 0.4 = 0.6 \end{aligned}$$
- $$\begin{aligned} E_2 &= \mu(\{w_1\}) \times (s_{21} - s_{22}) + \mu(\{w_1, w_2\}) \times (s_{22} - s_{23}) \\ &\quad + \mu(\{w_1, w_2, w_3\}) \times (s_{23} - s_{24}) + \mu(\{w_1, w_2, w_3, w_4\}) \times s_{24} \\ &= 0.4 \times 0.0 + 0.4 \times 0.0 + 0.4 \times 0.0 + 1.0 \times 0.8 = 0.8 \end{aligned}$$
- $$\begin{aligned} E_3 &= \mu(\{w_1\}) \times (s_{31} - s_{32}) + \mu(\{w_1, w_2\}) \times (s_{32} - s_{34}) \\ &\quad + \mu(\{w_1, w_2, w_4\}) \times (s_{34} - s_{33}) + \mu(\{w_1, w_2, w_3, w_4\}) \times s_{33} \\ &= 0.4 \times 0.2 + 0.4 \times 0.0 + 0.4 \times 0.2 + 1.0 \times 0.6 = 0.76 \end{aligned}$$
- $$\begin{aligned} E_4 &= \mu(\{w_3\}) \times (s_{43} - s_{44}) + \mu(\{w_3, w_4\}) \times (s_{44} - s_{42}) \\ &\quad + \mu(\{w_2, w_3, w_4\}) \times (s_{42} - s_{41}) + \mu(\{w_1, w_2, w_3, w_4\}) \times s_{41} \\ &= 0.2 \times 0.0 + 0.7 \times 0.2 + 0.7 \times 0.2 + 1.0 \times 0.4 = 0.68 \end{aligned}$$
- $$\begin{aligned} E_5 &= \mu(\{w_2\}) \times (s_{52} - s_{51}) + \mu(\{w_1, w_2\}) \times (s_{51} - s_{53}) \\ &\quad + \mu(\{w_1, w_2, w_3\}) \times (s_{53} - s_{54}) + \mu(\{w_1, w_2, w_3, w_4\}) \times s_{54} \\ &= 0.2 \times 0.2 + 0.4 \times 0.0 + 0.4 \times 0.2 + 1.0 \times 0.6 = 0.72 \end{aligned}$$

Step 3: Recommendation

The Choquet integral scores are:

$$E_1 = 0.6, \ E_2 = 0.8, \ E_3 = 0.76, \ E_4 = 0.68, \ E_5 = 0.72$$

Since Cup o_2 has the highest score of $E_2 = 0.8$, it is the most suitable recommendation.

6.3 Example Answers to Problems in Chapter 3

6.3.1 Example Answer to Problem 3.1: Education System

1. Interconnectedness:
   - Faculty-Student Relationships: Strong relationships foster student satisfaction, academic success, and research productivity.

- Curriculum and Pedagogy: The curriculum must align with industry and societal needs, while teaching methods should cultivate critical thinking and problem-solving skills.
- Infrastructure and Resources: High-quality facilities, such as state-of-the-art labs and digital libraries, are essential for effective learning and research.
- External Influences: Government regulations, global trends, and economic conditions influence institutional reputation, funding, and enrollment.

2. Holism:

- Student Experience: A well-rounded education integrates academics, research opportunities, social engagement, and career preparation, including internships or global exchange programs.
- Faculty Development: Supporting faculty with mentorship, professional development, and research resources enhances institutional effectiveness.
- Institutional Culture: Fostering an inclusive, innovative, diverse environment encourages collaboration and knowledge exchange.
- Strategic Planning: Aligning long-term strategies with the institution's mission and vision ensures sustainable growth.

3. Feedback Loops:

- Student Performance: Academic and research performance provides actionable insights for faculty and administrators.
- Faculty Evaluation: Reviews of teaching and research effectiveness inform professional development and curriculum adjustments.
- Institutional Assessment: Regular evaluations of programs and services ensure alignment with institutional goals.
- Reinforcing Feedback Loop: Investment in faculty research increases publications, such as papers in top-tier journals, attracts high-caliber students, and enhances institutional reputation.
- Balancing Feedback Loop: Declining student satisfaction could lower enrollment, reduce funding, and ultimately compromise program quality, creating a downward spiral unless corrective measures are implemented.

4. Emergent Properties:

- Innovative Research: Collaboration between faculty and students can lead to groundbreaking discoveries.
- Strong Graduate Network: A robust network of graduates provides mentorship, career opportunities, and financial support to the institution.
- Vibrant Academic Community: A thriving environment of intellectual exchange, such as through interdisciplinary research collaborations and student-led initiatives, fosters critical thinking, innovation, and academic excellence.

This systemic perspective highlights the intricate relationships between a graduate school and the broader education system. By understanding and addressing these dynamics, institutions can effectively navigate challenges and accomplish their objectives.

6.3.2 Example Answer to Problem 3.2: Causal Loop Diagram

We can consider the following two feedback loops:

1. Reinforcing Loop
   - Cycle: More bicycles available → More users → Reduced traffic congestion → Improved air quality and road safety → Increased public adoption of bike-sharing.
   - Impact: This loop promotes growth, as environmental and safety benefits encourage more users, particularly among younger and health-conscious demographics.
2. Balancing Loop
   - Cycle: Increased demand → Higher competition → Oversupply of bikes → Reduced utilization rates and parking challenges → Stricter government regulations → Hindered industry growth.
   - Impact: This loop highlights constraints such as improper parking, traffic issues, and regulatory burdens, which can slow the industry's expansion.

As shown in Fig. 6.1, a causal loop diagram can visually represent these relationships using the given variables.

From the figure, we can observe:

- Reinforcing Loop Realization: Over time, increased user adoption can lead to a shift away from traditional transportation modes such as taxis and buses.
- Balancing Loop Challenges: In the short term, oversupply, limited parking, and increased traffic accident rates (especially in cities lacking bike lanes) may require regulatory intervention.

A Complex System: By incorporating additional variables, a more nuanced causal loop diagram (Fig. 6.2) can be constructed, revealing bottlenecks and opportunities.

Potential bottlenecks include:

1. Limited Parking: Insufficient parking discourages users and creates urban clutter.
2. Intense Competition: An excessive number of market entrants drives down profits, making business survival challenging.
3. Strict Regulations: Compliance requirements increase operational costs and limit scalability.
4. Poor Urban Environment: High pollution levels and inadequate infrastructure deter cycling.

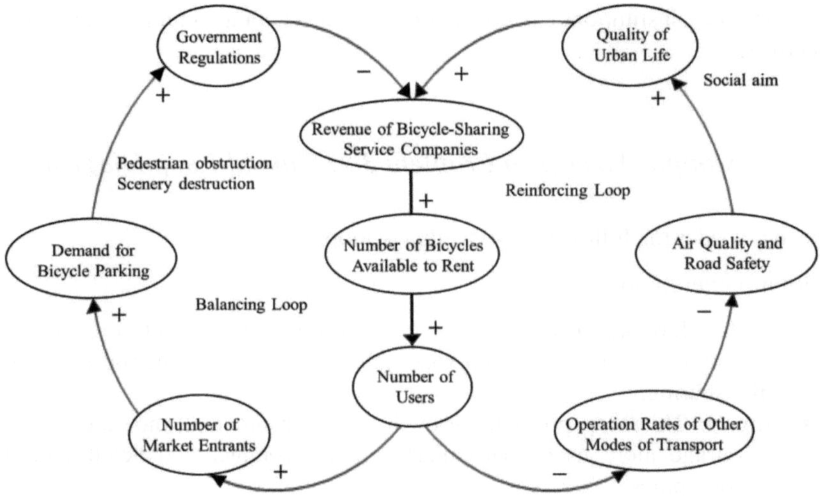

Fig. 6.1 A basic causal loop diagram representing the interactions within the bicycle-sharing business

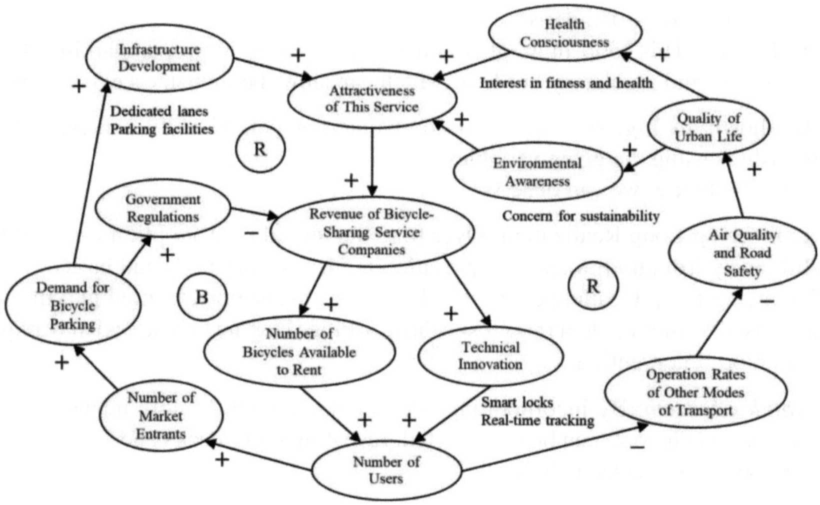

Fig. 6.2 A slightly more complex causal loop diagram revealing bottlenecks and opportunities

There are several opportunities for improvement:

1. Growing Environmental Awareness: Public concern for sustainability can boost demand for eco-friendly transport options.
2. Technological Innovation: Smart locks, real-time tracking, and app-based services enhance user experience and operational efficiency.

3. Urban Development Initiatives: Investments in bike-friendly infrastructure, such as dedicated lanes and parking facilities, can improve usability.
4. Health and Wellness Trends: Rising interest in fitness and health supports cycling as a preferred mode of transport.

Conclusion: Bicycle-sharing businesses operate within a complex, interconnected system influenced by both reinforcing and balancing loops. To ensure long-term success, these businesses must address key bottlenecks (e.g., parking, competition, and regulations) while leveraging environmental awareness, technological innovation, and infrastructure development opportunities. A well-designed causal loop diagram provides valuable insights into the challenges and growth strategies of this evolving industry.

6.3.3 Example Answer to Problem 3.3: Decision Tree

Following the problem statement, we will consider three water level conditions (Low, Normal, High), two wind strength categories (None/Weak, Strong), and two rainfall intensity levels (None/Light, Heavy). Since the notation becomes complex, at each step, divide a group into a left group and a right group, and denote the information gain ΔG as "Gain."

1. Root Gini Impurity
   - Entire dataset: 5 Yes and 7 No

$$G_{\text{root}} = 1 - \left(\frac{5}{12}\right)^2 - \left(\frac{7}{12}\right)^2 = \frac{70}{144} \approx 0.486$$

2. Splitting by Water Level

   A. Left group (Low): 2 Yes and 2 No; Right group (Normal + High): 3 Yes and 5 No
      - $G_{\text{left}} = 1 - \left(\frac{2}{4}\right)^2 - \left(\frac{2}{4}\right)^2 = \frac{8}{16} = 0.5$
      - $G_{\text{right}} = 1 - \left(\frac{3}{8}\right)^2 - \left(\frac{5}{8}\right)^2 = \frac{30}{64} \approx 0.469$
      - $\text{Gain} = G_{root} - \frac{4}{12} \times G_{\text{left}} - \frac{8}{12} \times G_{\text{right}} = \frac{70}{144} - \frac{4}{12} \times \frac{8}{16} - \frac{8}{12} \times \frac{30}{64} \approx 0.007$

   B. Left group (Normal): 2 Yes and 2 No; Right group (Low + High): 3 Yes and 5 No
      - The same Gini impurities for left and right groups as above.
      - Gain ≈ 0.007

   C. Left group (High): 1 Yes and 3 No; Right group (Low + Normal): 4 Yes and 4 No
      - $G_{\text{left}} = 1 - \left(\frac{1}{4}\right)^2 - \left(\frac{3}{4}\right)^2 = \frac{6}{16} = 0.375$

- $G_{\text{right}} = 1 - \left(\frac{4}{8}\right)^2 - \left(\frac{4}{8}\right)^2 = \frac{32}{64} = 0.5$
- $\text{Gain} = G_{\text{root}} - \frac{4}{12} \times G_{\text{left}} - \frac{8}{12} \times G_{\text{right}} = \frac{70}{144} - \frac{4}{12} \times \frac{6}{16} - \frac{8}{12} \times \frac{32}{64} \approx 0.028$

3. Splitting by Wind Strength

   - Left group (None/Weak): 4 Yes and 2 No; Right group (Strong): 1 Yes and 5 No
     - $G_{\text{left}} = 1 - \left(\frac{4}{6}\right)^2 - \left(\frac{2}{6}\right)^2 = \frac{16}{36} \approx 0.444$
     - $G_{\text{right}} = 1 - \left(\frac{1}{6}\right)^2 - \left(\frac{5}{6}\right)^2 = \frac{10}{36} \approx 0.278$
     - $\text{Gain} = G_{\text{root}} - \frac{6}{12} \times G_{\text{left}} - \frac{6}{12} \times G_{\text{right}} = \frac{70}{144} - \frac{6}{12} \times \frac{16}{36} - \frac{6}{12} \times \frac{10}{36} = 0.125$

4. Splitting by Rainfall

   - Left group (None/Light): 4 Yes and 2 No; Right group (Heavy): 1 Yes and 5 No
     - The same Gini impurities for left and right groups as above.
     - Gain = 0.125

5. Optimal Split

   - Since both Wind Strength and Rainfall yield the greatest information gain of 0.125, , either can be chosen as the first split. Let us split cases by Wind Strength and call the left (None/Weak) and right (Strong) groups $G1$ and $G2$, respectively. These groups comprise the following cases:

$$G1 = \{1, 2, 3, 8, 9, 12\}, G2 = \{4, 5, 6, 7, 10, 11\}$$

   - Gini impurities for $G1$ (4 Yes and 2 No) and $G2$ (1 Yes and 5 No) are:

$$G_{\text{G1}} = 1 - \left(\frac{4}{6}\right)^2 - \left(\frac{2}{6}\right)^2 = \frac{16}{36} \approx 0.444$$

$$G_{\text{G2}} = 1 - \left(\frac{1}{6}\right)^2 - \left(\frac{5}{6}\right)^2 = \frac{10}{36} \approx 0.278$$

6. Splitting G1 by Water Level

   A. Left group (Low): 2 Yes and 0 No; Right group (Normal + High): 2 Yes and 2 No

   - $G_{\text{left}} = 1 - \left(\frac{2}{2}\right)^2 - \left(\frac{0}{2}\right)^2 = \frac{0}{4} = 0$
   - $G_{\text{right}} = 1 - \left(\frac{2}{4}\right)^2 - \left(\frac{2}{4}\right)^2 = \frac{8}{16} = 0.5$
   - $\text{Gain} = G_{\text{G1}} - \frac{2}{6} \times G_{\text{left}} - \frac{4}{6} \times G_{\text{right}} = \frac{16}{36} - \frac{2}{6} \times \frac{0}{4} - \frac{4}{6} \times \frac{8}{16} \approx 0.111$

B. Left group (Normal): 1 Yes and 1 No; Right group (Low + High): 3 Yes and 1 No

- $G_{\text{left}} = 1 - \left(\frac{1}{2}\right)^2 - \left(\frac{1}{2}\right)^2 = \frac{2}{4} = 0.5$
- $G_{\text{right}} = 1 - \left(\frac{3}{4}\right)^2 - \left(\frac{1}{4}\right)^2 = \frac{6}{16} = 0.375$
- $\text{Gain} = G_{\text{G1}} - \frac{2}{6} \times G_{\text{left}} - \frac{4}{6} \times G_{\text{right}} = \frac{16}{36} - \frac{2}{6} \times \frac{2}{4} - \frac{4}{6} \times \frac{6}{16} \approx 0.028$

C. Left group (High): 1 Yes and 1 No; Right group (Low + Normal): 3 Yes and 1 No

- The same Gini impurities for left and right groups as above.
- Gain ≈ 0.028

7. Splitting G1 by Rainfall:

- Left group (None/Light): 3 Yes and 0 No; Right group (Heavy): 1 Yes and 2 No
  - $G_{\text{left}} = 1 - \left(\frac{3}{3}\right)^2 - \left(\frac{0}{3}\right)^2 = \frac{0}{9} = 0$
  - $G_{\text{right}} = 1 - \left(\frac{1}{3}\right)^2 - \left(\frac{2}{3}\right)^2 = \frac{4}{9} \approx 0.444$
  - $\text{Gain} = G_{\text{G1}} - \frac{3}{6} \times G_{\text{left}} - \frac{3}{6} \times G_{\text{right}} = \frac{16}{36} - \frac{3}{6} \times \frac{0}{9} - \frac{3}{6} \times \frac{4}{9} \approx 0.222$

8. Optimal Split of $G1$

- Partitioning G1 based on Rainfall yields the highest gain, so dividing into G11 (None/Light) and G12 (Heavy) is appropriate:

$$G11 = \{2, 8, 12\}, G12 = \{1, 3, 9\}$$

- As all cases in $G11$ exhibit a "Yes" outcome for Operation, further partitioning is unnecessary. $G12$ can be subdivided based on Water Level, resulting in:
  - $G121 = \{3\}$: Water Level = Low, Operation = Yes
  - $G122 = \{1, 9\}$: Water Level = Normal or High, Operation = No

9. Splitting $G2$ by Water Level

A. Left group (Low): 0 Yes and 2 No; Right group (Normal + High): 1 Yes and 3 No

- $G_{\text{left}} = 1 - \left(\frac{0}{2}\right)^2 - \left(\frac{2}{2}\right)^2 = \frac{0}{4} = 0$
- $G_{\text{right}} = 1 - \left(\frac{1}{4}\right)^2 - \left(\frac{3}{4}\right)^2 = \frac{6}{16} = 0.375$
- $\text{Gain} = G_{\text{G2}} - \frac{2}{6} \times G_{\text{left}} - \frac{4}{6} \times G_{\text{right}} = \frac{10}{36} - \frac{2}{6} \times \frac{0}{4} - \frac{4}{6} \times \frac{6}{16} \approx 0.028$

B. Left group (Normal): 1 Yes and 1 No; Right group (Low + High): 0 Yes and 4 No

- $G_{\text{left}} = 1 - \left(\frac{1}{2}\right)^2 - \left(\frac{1}{2}\right)^2 = \frac{2}{4} = 0.5$
- $G_{\text{right}} = 1 - \left(\frac{0}{4}\right)^2 - \left(\frac{4}{4}\right)^2 = \frac{0}{16} = 0$

  - Gain $= G_{\text{G2}} - \frac{2}{6} \times G_{\text{left}} - \frac{4}{6} \times G_{\text{right}} = \frac{10}{36} - \frac{2}{6} \times \frac{2}{4} - \frac{4}{6} \times \frac{0}{16} \approx 0.111$

C. Left group (High): 0 Yes and 2 No; Right group (Low + Normal): 1 Yes and 3 No

  - The same Gini impurities for left and right groups as in Case A above.
  - Gain ≈ 0.028

10. Splitting $G2$ by Rainfall

  - Left group (None/Light): 1 Yes and 2 No; Right group (Heavy): 0 Yes and 3 No
    - $G_{\text{left}} = 1 - \left(\frac{1}{3}\right)^2 - \left(\frac{2}{3}\right)^2 = \frac{4}{9} \approx 0.444$
    - $G_{\text{right}} = 1 - \left(\frac{0}{3}\right)^2 - \left(\frac{3}{3}\right)^2 = \frac{0}{9} = 0$
    - Gain $= G_{\text{G2}} - \frac{3}{6} \times G_{\text{left}} - \frac{3}{6} \times G_{\text{right}} = \frac{10}{36} - \frac{3}{6} \times \frac{4}{9} - \frac{3}{6} \times \frac{0}{9} \approx 0.056$

11. Optimal Split of $G2$

  - Dividing $G2$ by Water Level (Case B) yields the largest gain, so dividing into $G21$ (Normal) and $G22$ (Low + High) is appropriate:

$$G21 = \{4, 5\}, G22 = \{6, 7, 10, 11\}$$

  - Since all cases in $G22$ have a "No" outcome for Operation, no further division is necessary. $G21$, however, can be further partitioned by Rainfall, creating two new groups:
    - $G211 = \{5\}$: Rainfall = None/Light, Operation = Yes
    - $G212 = \{4\}$: Rainfall = Heavy, Operation = No

12. Fig. 6.3 illustrates the decision tree for determining whether a rafting trip should proceed.
13. Decision Rules for Service Operation

I. Rafting will proceed if:

  - The Wind Strength is None/Weak, and the Rainfall is None/Light.
  - The Wind Strength is None/Weak, and the Water Level is Low.
  - The Water Level is Normal, and the Rainfall is None/Light.

II. Rafting will be canceled if:

  - The Wind Strength is Strong, and the Water Level is Low or High.
  - The Rainfall is Heavy, and the Water Level is Normal or High.
  - The Wind Strength is Strong, and the Rainfall is Heavy.

Note: This dataset is small, so these rules may not generalize well to other cases. In practice, it is crucial to train models on larger datasets and validate their performance on separate data not used for training.

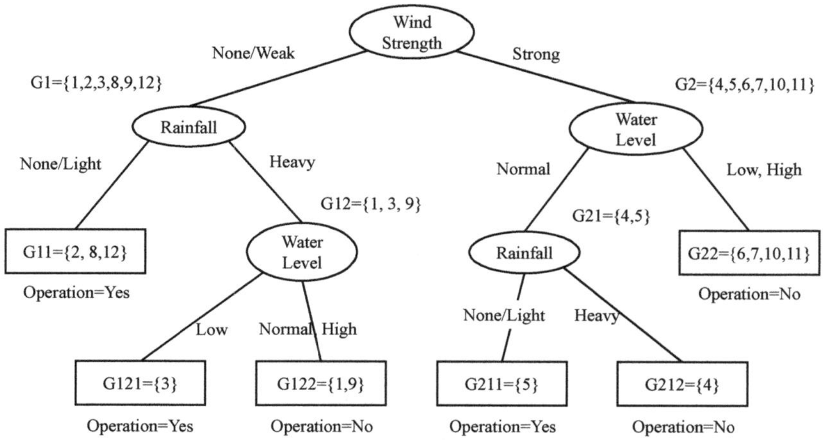

Fig. 6.3 A decision tree for deciding whether to go whitewater rafting

6.3.4 *Example Answer to Problem 3.4: Bayesian Decision Theory*

We define the following:

- H_1: Hypothesis that you are infected
- H_2: Hypothesis that you are not infected
- D: Positive test result

Bayes' Theorem states:

$$P(H_i|D) = \frac{P(D|H_i)P(H_i)}{P(D)}, \quad i = 1,2$$

where $P(D)$ is the total probability of a positive test result, given by:

$$P(D) = P(D \cap H_1) + P(D \cap H_2) = P(D|H_1)P(H_1) + P(D|H_2)P(H_2)$$

The test characteristics are given as:

- $P(D|H_1) = 0.7$ (sensitivity)
- $P(D|H_2) = 0.1$ (false positive rate)

Case 1: Asymptomatic Testing

If you are asymptomatic, the probability of being infected is the city's infection rate:

$$P(H_1) = 0.03, \; P(H_2) = 0.97$$

Using Bayes' Theorem, the posterior probability of being infected is:

$$P(H_1|D) = \frac{P(D|H_1)P(H_1)}{P(D)} = \frac{P(D|H_1)P(H_1)}{P(D|H_1)P(H_1) + P(D|H_2)P(H_2)}$$

Substituting the values:

$$P(H_1|D) = \frac{0.7 \times 0.03}{0.7 \times 0.03 + 0.1 \times 0.97} \approx 0.178$$

The probability of not being infected is:

$$P(H_2|D) = 1 - P(H_1|D) \approx 0.822$$

Interpretation: If you test positive but have no symptoms, there is approximately a 17.8% chance that you are actually infected.

Case 2: Symptomatic Testing

If you have symptoms and estimate an 80% chance of infection:

$$P(H_1) = 0.8, \; P(H_2) = 0.2$$

The posterior probability of being infected is:

$$P(H_1|D) = \frac{P(D|H_1)P(H_1)}{P(D)} = \frac{P(D|H_1)P(H_1)}{P(D|H_1)P(H_1) + P(D|H_2)P(H_2)}$$

Substituting the values:

$$P(H_1|D) = \frac{0.7 \times 0.8}{0.7 \times 0.8 + 0.1 \times 0.2} \approx 0.966$$

The probability of not being infected is:

$$P(H_2|D) = 1 - P(H_1|D) \approx 0.034$$

Interpretation: If you have symptoms and your test is positive, there is approximately a 96.6% chance that you are actually infected.

Conclusion: Bayes' Theorem highlights the importance of considering prior probabilities:

- In the asymptomatic case, a positive test result is far less conclusive due to the low baseline infection rate.
- In the symptomatic case, the high prior probability of infection significantly increases the likelihood of a true positive.

This approach demonstrates how diagnostic test results should be interpreted in context with other information, such as symptoms and prevalence rates.

6.3.5 Example Answer to Problem 3.5: Soft Systems Methodology

Stage 3—Root Definition: The remote health management system aims to:

- Improve the health and well-being of individuals with mild diabetes.
- Alleviate the workload on healthcare facilities.
- Reduce healthcare costs for the local government.
- Encourage collaboration among local businesses to drive health innovation.
- Enhance residents' quality of life and provide a scalable model for managing additional health conditions in the future.

Stage 4—Conceptual Model: Key problems and issues include:

1. Patient Engagement: How can we ensure that patients follow recommended health practices (e.g., diet, exercise, and self-monitoring)?
2. Technology Integration: What devices and software are needed to ensure seamless data collection and sharing?
3. Data Utilization: How can collected data (e.g., blood glucose levels and meal logs) be leveraged to deliver personalized recommendations?
4. Financial Planning: How can the system remain affordable for both the city and participants?
5. Collaboration: How can stakeholders (e.g., hospitals and tech providers) work together effectively?

Note: As shown in Fig. 3.15, these issues and their interrelationships can be visualized as a conceptual model. Rather than framing them as questions, focused instructions such as "what to look for" or "what to consider" may be more actionable.

Stage 5—Potential Solutions: Examples of potential solutions are summarized in Table 6.11.

Stage 6—Systemically Desirable and Culturally Feasible Solutions: To implement the system effectively, the following action plan is proposed:

- Pilot Program: Conduct a one-year pilot program funded by national and local governments.
- City Oversight: The city government will manage and oversee the program's execution.
- Technology Procurement:
  - Purchase biosensors to monitor patient health metrics.
  - Develop a user-friendly app for real-time data tracking and communication.

Table 6.11 Examples of potential solutions

Challenge	Potential solution
Patient engagement	Provide regular advice on meal planning, exercise, and self-monitoring. Schedule occasional in-person doctor visits
Technology integration	Procure biosensors and commission an IT company to develop a mobile app for data sharing and feedback
Data utilization	Collect patient data (e.g., blood glucose levels, meal photos) and use it to provide personalized advice via dietitians
Financial planning	Initiate a government-subsidized pilot program to evaluate feasibility and impact before full-scale implementation
Collaboration	Organize regular meetings between hospitals, sensor companies, and software developers to ensure alignment

- Expert Support: Engage dietitians or medical professionals to provide daily guidance to participants.
- Patient Recruitment: Enroll approximately 60 patients with mild diabetes from local hospitals.
- Patient Cooperation: Motivate participants to actively share health data and adhere to prescribed guidance.

In addition to these steps, it is essential to consider the ethical implications of the proposed solutions to ensure their fairness and acceptability. Key ethical considerations include:

- Ensuring patient data privacy and security to protect sensitive information.
- Promoting inclusivity by making the system affordable and accessible to all residents.
- Continuously assessing the system's effectiveness in improving health outcomes and societal well-being.

By addressing these ethical concerns alongside technical and operational challenges, the remote health management system can achieve its goals while maintaining public trust and support.

Conclusion: The remote health management system presents a promising solution to rising healthcare costs and challenges associated with an aging population. By aligning stakeholder interests, leveraging technology, and fostering collaboration, this approach can significantly improve public health outcomes while ensuring feasibility and cultural acceptability. A carefully structured pilot program will serve as a foundation for sustainable, long-term implementation.

6.4 Example Answers to Problems in Chapter 4

6.4.1 Example Answer to Problem 4.1: Reasoning Methods

Selecting lunch can be surprisingly tricky. While we often rely on gut feelings, applying different decision-making approaches can help us make more informed choices.

I. Inductive Reasoning: Learning from Experience

- Drawing from Experience: If you have enjoyed a dish at a restaurant before, you might assume that ordering it again will also be a good choice.
- Observational Learning: Seeing many customers order a particular dish could indicate that it is popular and tasty.

II. Deductive Reasoning: Using Rules and Logic

- Making Decisions Based on Rules: If you follow a specific diet (e.g., vegetarian, gluten-free, or low-carb), you can narrow down your choices by selecting dishes that meet those needs.
- Process of Elimination: If you do not like spicy food or prefer seafood, you can rule out dishes that do not match your preferences.

III. Abductive Reasoning: Making an Educated Guess

- Inferring from Clues: Words like "chef's special" or "homemade" might suggest that the dish is unique and full of flavor.
- Exploring New Options: Choosing an unfamiliar dish based on an intriguing name or description can lead to an enjoyable discovery.

IV. Integrating Different Reasoning Methods

In reality, we often use a mix of these reasoning methods:

- Inductive: You remember liking a specific type of food before.
- Deductive: You eliminate options that do not fit your diet or preferences.
- Abductive: You take a chance on a dish that sounds exciting.

By applying these reasoning methods, we can turn a simple lunch decision into a more satisfying and enjoyable experience.

6.4.2 Example Answer to Problem 4.2: Knowledge Management

1. The Challenges of Harnessing Tacit Knowledge

- *Articulation*: Tacit knowledge is personal, abstract, and often subconscious, making it difficult to express or document. Different perspectives can lead to misinterpretations or communication barriers.

- *Transfer*: Traditional knowledge-sharing methods often fail to capture tacit knowledge, which requires personal interaction, observation, and mentorship.
- *Dependency on Key Individuals*: Tacit knowledge is often concentrated within a few individuals, posing a significant risk of loss when they leave the organization.
- *Preservation*: Tacit knowledge is context-dependent and fleeting. Resistance to documentation due to time constraints, intellectual property concerns, or a preference for informal sharing further complicates preservation efforts.

2. Strategies for Leveraging Tacit Knowledge

   - Cultivate a Knowledge-Sharing Culture
     - Build a supportive environment where employees feel comfortable sharing experiences.
     - Encourage informal interactions through mentoring, coaching, and cross-functional collaboration.
   - Leverage Storytelling and Case Studies
     - Use stories, anecdotes, and real-world examples to convey complex concepts and best practices.
     - Highlight case studies that illustrate lessons learned and successes.
   - Create Communities of Practice:
     - Establish groups where employees with shared interests can connect, exchange ideas, and learn from each other.
   - Implement Knowledge Management Systems:
     - Develop intuitive tools for capturing, storing, and retrieving tacit knowledge.
     - Focus on making documentation easy and meaningful.
   - Invest in Experiential Learning:
     - Facilitate hands-on learning through observation, shadowing, and job rotations.
     - Promote apprenticeship programs to transfer expertise effectively.

Conclusion: By tackling the challenges of tacit knowledge and applying these strategies, organizations can unlock their full potential, drive innovation, and enhance decision-making. Harnessing tacit knowledge strengthens organizational performance and creates a sustainable competitive advantage in a rapidly evolving business landscape.

6.4.3 Example Answer to Problem 4.3: Logic Tree

1. *Drawing a Logic Tree*: Following the instructions, I have drawn a logic tree for my job search, as shown in Fig. 6.4, as a computer science student.
2. *Imagining a Career Path*: Based on the logic tree, I have outlined a potential career path:

   - Industry: Technology—Software
     - Software development aligns with the core principles of a Computer Science education.
   - Role: Software Engineer—Backend
     - I am especially interested in backend development because I am passionate about data structures, algorithms, and server-side logic.
   - Work Environment: Startup—Growth-Stage
     - I prefer the dynamic environment of a startup, where I can significantly impact the company's growth.
   - Lifestyle: Fast-Paced - High-Growth
     - I will embrace a fast-paced lifestyle to learn quickly, grow professionally, and contribute to a thriving organization.

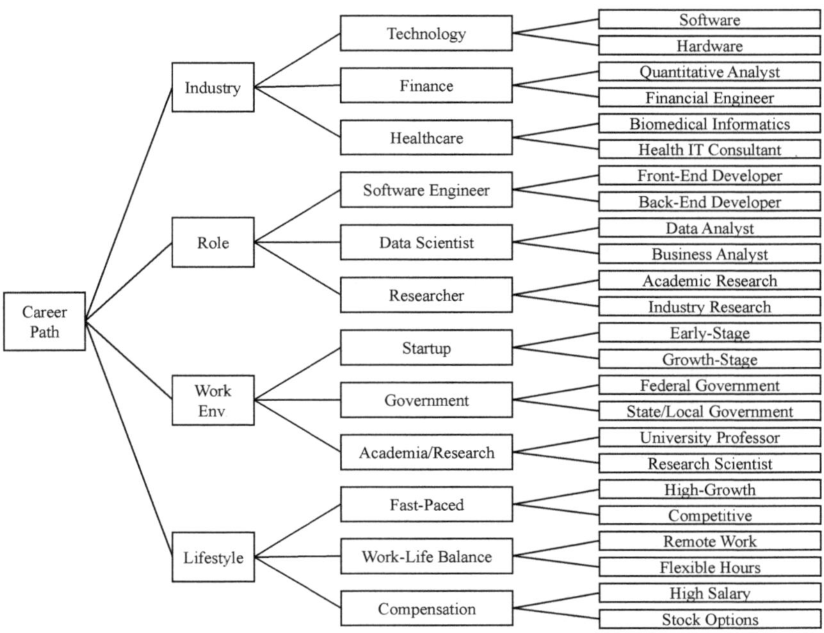

Fig. 6.4 An example logic tree illustrating the career decision process of a computer science student

This career path would begin with a role as a backend engineer at a growing tech startup. I would focus on building expertise in backend frameworks (e.g., Node.js, Python/Django), databases, and cloud computing. Over time, I would aim to advance into senior technical roles or explore leadership opportunities.

Note: The above is just one example of a possible career path. The power of the logic tree lies in its adaptability. It enables you to explore and tailor various options to your unique career aspirations. By reflecting on your interests, values, and skills, you can craft a career path that aligns with your vision for the future.

6.4.4 Example Answer to Problem 4.4: SWOT and Cross-SWOT Analysis

Cross-SWOT analysis for summer menu planning is performed as follows:

1. Leveraging Strengths to Seize Opportunities:
   - (Strength: Brand Recognition) × (Opportunity: Food Trends) → Introduce limited-time, trend-inspired dishes (e.g., plant-based sushi rolls or fusion flavors) to engage loyal customers and attract new patrons.
   - (Strength: Efficient Operations) × (Opportunity: Seasonal Ingredients) → Quickly adapt the menu to incorporate fresh, seasonal ingredients while maintaining operational efficiency.
   - (Strength: Fresh Ingredients) × (Opportunity: Online Ordering and Delivery) → Offer premium, high-quality sushi options for delivery, reinforcing the restaurant's reputation for freshness.
2. Minimizing Threats Using Strengths:
   - (Threat: Economic Downturn) × (Strength: Brand Recognition) → Focus marketing efforts on loyal customer segments by offering value-driven promotions.
   - (Threat: Increased Competition) × (Strength: Efficient Operations) → Optimize costs through operational efficiency, enabling competitive pricing without sacrificing quality.
   - (Threat: Supply Chain Disruptions) × (Strength: Fresh Ingredients) → Strengthen supplier relationships and establish contingency plans for ingredient sourcing.
3. Mitigating Weaknesses Through Opportunities:
   - (Weakness: Limited Menu Variety) × (Opportunity: Food Trends) → Expand variety by creating themed seasonal promotions (e.g., summer-inspired sushi platters).
   - (Weakness: Seasonal Fluctuations) × (Opportunity: Online Ordering and Delivery) → Launch exclusive online promotions or delivery discounts during off-peak times to boost demand.

- (Weakness: Price Sensitivity) × (Opportunity: Food Trends) → Introduce value-added, premium options, such as curated "chef's specials," to target less price-sensitive customers.

4. Addressing Weaknesses and Threats Comprehensively:

   - (Weakness: Limited Menu Variety) × (Threat: Increased Competition) → Invest in menu innovation through R&D to stay ahead of competitors with unique offerings.
   - (Weakness: Price Sensitivity) × (Threat: Economic Downturn) → Adopt a tiered pricing model to appeal to both budget-conscious and premium customers.
   - (Weakness: Seasonal Fluctuations) × (Threat: Supply Chain Disruptions) → Develop a flexible sourcing strategy to minimize the effects of seasonal variations and supply chain issues.

The strategies from the Cross-SWOT analysis can be classified into the following categories:

1. Differentiation Strategy (Creating unique value to stand out from competitors)

   - Introducing limited-time trend-inspired dishes (e.g., plant-based sushi rolls, fusion flavors)
   - Expanding high-quality delivery options (premium sushi for online ordering)
   - Differentiating through unique menu innovations (seasonal-themed promotions like summer-inspired sushi platters)
   - Investing in menu R&D to stay ahead of competitors

2. Growth Strategy (Expanding market presence and increasing revenue)

   - Efficiently adapting the menu to seasonal ingredients (leveraging fresh ingredients for summer offerings)
   - Boosting online sales during off-peak times (exclusive online promotions and delivery discounts)
   - Strengthening promotions for loyal customers (value-driven campaigns to retain and grow the customer base)
   - Implementing a flexible pricing strategy (tiered pricing to attract both budget-conscious and premium customers)

3. Risk Mitigation Strategy (Addressing external threats and internal weaknesses)

   - Optimizing costs to maintain competitive pricing (efficiency-driven cost management)
   - Strengthening supplier relationships and contingency planning (mitigating supply chain disruptions)
   - Developing a flexible sourcing strategy (adapting to seasonal fluctuations and supply risks)

These strategies will help optimize the summer menu planning, making the most of market opportunities while mitigating risks.

6.4.5 Example Answer to Problem 4.5: Innovation Theory

1. Sustaining Innovation or Disruptive Innovation?

   - Amazon is a prime example of a company that transitioned from disruptive to sustaining innovation over time.
   - *Disruptive Innovation*: Amazon revolutionized the retail industry by introducing a highly convenient and efficient online shopping experience. It was a classic case of disruptive innovation, as it challenged and ultimately transformed the established brick-and-mortar retail model. Amazon fundamentally changed how customers interact with retail by eliminating physical storefronts, offering competitive prices, and leveraging technology to streamline shopping.
   - *Sustaining Innovation*: As Amazon expanded and matured, its focus shifted to sustaining innovation, marked by continuous refinement and enhancement of its business model. Examples of this include:

     – Expanding Product Lines: Diversifying from books to electronics, groceries, apparel, and more.
     – Enhancing Logistics: Utilizing cutting-edge technologies such as AI, drones, and robotics to optimize its delivery network.
     – Improving Customer Experience: Offering faster shipping (e.g., Amazon Prime), personalized recommendations, and exceptional customer service.

While Amazon continues to innovate groundbreakingly in some areas (e.g., cloud computing and AI), its core e-commerce business primarily focuses on incremental improvements and operational excellence, which aligns with sustaining innovation.

2. Value Innovation and Blue Ocean Strategy:

   - Amazon epitomizes value innovation and strongly aligns with the principles of Blue Ocean Strategy.
   - *Value Innovation:* Amazon created a new market space by simultaneously pursuing cost leadership and differentiation, offering unmatched value to customers. Amazon's strategic moves can be broken down as follows:

     – Eliminating: Did away with physical stores, drastically cutting operational costs and transferring the savings to customers.
     – Reducing: Lowered prices for many products by leveraging supply chain efficiency and economies of scale.
     – Raising: Enhanced convenience and selection by providing a vast product assortment, fast shipping (e.g., Amazon Prime), and personalized recommendations.
     – Creating: Pioneered a new market for online retail by combining convenience, affordability, and a seamless shopping experience.

- *Blue Ocean Strategy*: Amazon initially entered an underserved market, forging a new "blue ocean" by redefining the shopping experience. Key aspects of its Blue Ocean Strategy include:
  - Differentiation: Offering features such as one-click ordering, Prime delivery, and an extensive product selection unavailable in traditional retail.
  - Cost Leadership: Achieving cost advantages through supply chain optimization, automation, and economies of scale.

Through these strategies, Amazon bypassed traditional retail competition and established a dominant position in the online retail landscape. Its continuous expansion into cloud computing (AWS) and digital streaming further demonstrates its ability to create new markets and value propositions.

Conclusion: Amazon's evolution showcases a smooth shift from disruptive to sustaining innovation while exemplifying value innovation and Blue Ocean Strategy. By consistently redefining markets and enhancing customer value, Amazon remains a global leader in innovation across diverse industries.

6.5 Example Answers to Problems in Chapter 5

6.5.1 Example Answer to Problem 5.1: Synthetic Thinking

1. Differences Between Divergent-Convergent Thinking and Synthetic Thinking
   - Focus:
     A. Divergent-Convergent Thinking: Generates and refines ideas within a specific domain.
     B. Synthetic Thinking: Integrates knowledge from various domains to create innovative solutions and fresh insights.
   - Process:
     A. Divergent-Convergent Thinking: Follows a structured sequence—first generating multiple ideas (divergent), then evaluating and refining them to select the best options (convergent).
     B. Synthetic Thinking: Uses an interconnected and holistic approach, gathering, integrating, and reinterpreting information from diverse sources to develop new concepts.
2. Application to Planning a Summer Menu for a Sushi Restaurant
   A. Divergent-Convergent Thinking Approach
      i. Divergent Thinking (Idea Generation): Brainstorm various summer-themed sushi dishes, such as:
         - Spicy Tuna Roll with Citrus Glaze
         - Mango Avocado Roll with Yuzu Dressing
         - Chilled Udon with Tempura

ii. Convergent Thinking (Idea Evaluation): Assess these options based on:
    - Feasibility: Ingredient availability and preparation complexity.
    - Customer Appeal: Taste preferences and visual presentation.
    - Profitability: Cost of ingredients and pricing potential.

iii. Final Selection: Choose the most viable dishes and refine their presentation.

B. Synthetic Thinking Approach

- Gather Diverse Knowledge Sources:
  - Customer Data: Analyze past summer sales, customer feedback, and demographic trends.
  - Seasonal Trends: Research locally available summer seafood and produce.
  - Competitor Analysis: Examine how other sushi restaurants adapt their summer menus.
  - Cultural Influences: Explore traditional Japanese summer dishes and emerging culinary trends.
- Synthesize Information to Create a Unique Menu Concept:
  - Example 1: Combining customer data with seasonal trends might inspire a *Summer Bounty Roll* featuring locally sourced seasonal ingredients like grilled eel, shiso leaves, and pickled summer vegetables.
  - Example 2: Integrating cultural influences with competitor insights could result in a *Refreshing Citrus Sushi Platter*, incorporating yuzu-infused sauces, cold soba, and light seafood pairings.

Summary

- Divergent-Convergent Thinking:
  - Ideas are generated, refined, and selected within a structured framework.
  - Divergent thinking seeks to present possible solutions, so the range of ideas tends to be limited to the sushi menu.
- Synthetic Thinking:
  - Extends beyond structured problem-solving by integrating diverse knowledge sources.
  - Incorporates insights from other cultures, market trends, and customer data to develop new ideas.

6.5.2 Example Answer to Problem 5.2: AlphaGo

Both professional Go players and AI systems like AlphaGo strive for victory on the Go board, yet they employ distinct approaches and exhibit unique strengths.

Human Go players rely on:

- Intuition and Creativity: Human players make intuitive leaps, devising novel strategies and unexpected moves that challenge conventional wisdom.
- Contextual Understanding: They possess a deep grasp of the game's nuances, factoring in their opponent's playing style, game stage, and the broader strategic landscape.
- Learning from Experience: Human players refine their strategies over time by learning from victories, defeats, and mentorship, developing unique playing styles.
- Emotional Intelligence: While emotions can sometimes hinder performance, they also fuel determination, focus, and resilience.

AlphaGo, on the other hand, operates based on:

- Computational Power: AlphaGo can analyze millions of possible moves per second, evaluating board positions with exceptional speed and precision.
- Data-Driven Learning: It learns from extensive datasets of human games and, through self-play, identifies intricate patterns and strategies beyond human comprehension.
- Statistical Optimization: AlphaGo relies on probability calculations and statistical models to determine the optimal move in any given situation.
- Objectivity: Free from emotions and biases, AlphaGo makes decisions based solely on data-driven probability assessments.

Table 6.12 summarizes the differences.

Conclusion: While AlphaGo has achieved unparalleled success in Go, surpassing human champions in terms of speed and computational power, it is essential to recognize the unique strengths of human players. Creativity, intuition, and emotional intelligence remain invaluable aspects of the game, enriching the human experience and fostering a deeper understanding of its complexities. The ongoing interaction between human ingenuity and artificial intelligence in Go continues to be a

Table 6.12 Key differences between human professional Go players and AlphaGo

Aspect	Human Go players	AlphaGo
Learning mechanisms	Learn through experience, intuition, and mentorship	Learns through data analysis, self-play, and algorithmic refinement
Decision-making process	Rely on intuition, experience, and strategic thinking	Uses statistical analysis and pattern recognition
Creativity and innovation	Develop original strategies and innovative concepts	Displays creativity within learned patterns but lacks true conceptual innovation
Emotional intelligence	Operate with motivation, resilience, and adaptability	Functions without emotions, making decisions purely based on probabilities

compelling area of exploration, pushing the boundaries of both human and machine intelligence.

6.5.3 Example Answer to Problem 5.3: Tacit Integration

Humans recognize faces effortlessly, without consciously analyzing individual features like eye color or nose shape. Instead, we perceive the face holistically, integrating visual cues and past experiences in a seamless, subconscious process—an ability Polanyi refers to as tacit integration.

In contrast, AI-based recognition decomposes faces into discrete features and processes them algorithmically. While AI has advanced significantly, it still struggles to replicate the intuitive and adaptive nature of human perception.

The main differences between human and AI facial recognition are as follows.

Human facial recognition operates based on:

- Holistic Processing: Faces are perceived as a whole rather than as individual features.
- Contextual Awareness: Recognition is influenced by hairstyle, clothing, and surroundings.
- Subconscious Speed: Humans identify faces rapidly and intuitively.
- Adaptability: Recognition remains accurate despite aging, lighting changes, or partial obstructions.
- Emotional Intelligence: Humans interpret emotions and subtle expressions with ease.

On the other hand, AI facial recognition relies on:

- Feature-Based Analysis: Faces are broken down into distinct components (e.g., eyes, nose, mouth) for comparison.
- Data-Driven Learning: Recognition improves through exposure to large datasets.
- Computational Power: AI processes vast amounts of facial data in real time.
- Controlled Accuracy: AI performs best in consistent conditions but struggles with occlusions or variations.
- Limited Emotional Intelligence: AI is still developing the ability to interpret emotions.

While AI excels at speed and large-scale facial recognition, human recognition is more adaptable, intuitive, and context-aware. Tacit integration allows humans to recognize faces effortlessly in complex scenarios, a skill AI has yet to replicate fully.

6.5.4 *Example Answer to Problem 5.4: Knowledge Triad Synthesis*

Refining Rational Ideas by Incorporating Social Considerations: Certain types of sushi, particularly those made with seasonal winter fish, experience lower sales in summer due to opportunity loss rather than reduced customer demand. This insight leads to key menu optimization strategies:

- Emphasizing Seafood Demand: Capitalizing on the strong demand for seafood, including winter fish, even during summer, enhances the menu appeal. While importing winter fish may increase costs, it can also differentiate the restaurant and attract customers looking for premium options.
- Diversifying the Side Menu: Offering seasonal vegetable salads and tempura during warmer months aligns with customer preferences and provides greater variety.
- Aligning with Customer Expectations: Customers value fresh, seasonal, and reasonably priced fish. Balancing quality and affordability is essential for maintaining customer satisfaction.

With these refinements, systemically desirable ideas emerge:

- **Systemically Desirable Ideas:** To meet the continued demand for winter fish in summer, we strategically import select winter fish, despite slightly higher costs, to enhance menu appeal and profitability. At the same time, we prioritize fresh, seasonal, and reasonably priced fish to align with customer expectations. A varied side menu featuring light vegetable salads and seasonal tempura ensures diverse options that cater to summer dining preferences, making the menu more dynamic and appealing.

Refining Intuitive Ideas by Incorporating Social Considerations: We can enhance intuitive ideas by integrating customer expectations and social factors:

- Customer-Centric Approach: Understanding customer preferences for fresh, seasonal, and affordable options allows for a tailored menu that meets expectations.
- Enhancing Menu Diversity: Expanding the menu with seasonal salads and tempura enhances its appeal to a broader range of tastes and dietary needs.
- Family-Friendly Menus: Introducing affordable, kid-friendly sushi sets with familiar ingredients (e.g., omelet and chicken) attracts families, particularly during summer vacations.

These refinements lead to culturally feasible ideas that merge innovation with practicality:

- **Culturally Feasible Ideas:** By introducing special children's sushi sets featuring recognizable ingredients, the restaurant caters to young diners while maintaining affordability. Strategic sourcing, including imports and frozen options, ensures a consistent menu without compromising quality. Additionally, bulk purchasing of fresh fish during peak seasons helps control costs and improve profitability, creating a sustainable model for menu development.

Synthesizing Systemically Desirable and Culturally Feasible Ideas: By integrating the critical elements—customer preferences, reasonably priced fish, menu variety, kid-friendly options, and strategic sourcing—we develop a well-rounded summer menu strategy:

- **Synthesized Ideas:** By strategically balancing seasonal offerings with customer preferences, we create a consistent year-round dining experience. Fresh seasonal produce is highlighted, with light salads and summer desserts introduced during warmer months. Popular fish are sourced through fresh, imported, and frozen options to maintain a reliable and high-quality menu. Affordable, family-friendly options attract a broader customer base, ensuring that the restaurant remains a top choice for regular and seasonal diners. A harmonized approach that blends innovation, quality, and value establishes long-term success while maintaining a competitive edge in the market.

Final Thoughts: This synthesis ensures that the summer menu remains attractive, diverse, and profitable by aligning rational insights, intuitive creativity, and customer-driven expectations. The restaurant can optimize its summer menu by strategically sourcing ingredients, diversifying offerings, and catering to family dining trends while preserving customer satisfaction and business sustainability.

6.5.5 Example Answer to Problem 5.5: Strategy Story

For example, consider exploring research activities that promote knowledge synthesis.

I. **Stage 1—Intervention:**

1. Define the Problem: Planning research on knowledge synthesis enablers.
2. Develop a Knowledge Acquisition Plan:
   - Hypothesis Setting: Based on the Knowledge Construction Model, formulate a hypothesis about research activities contributing to knowledge synthesis as a research outcome.
   - Research method: Design a questionnaire survey to collect data and validate the hypothesis.
3. Research questions:
   - What research activities contribute to synthesizing knowledge as an outcome?
   - How do these activities vary across different academic fields?

II. **Stage 2—Imagination, Intelligence, Involvement:**

A. **Intention Story**—*Leveraging Personal Passions and Values* (Imagination): Many researchers seek to understand the key research practices that facilitate knowledge synthesis and improve research outcomes. To explore this, I aim to develop a hypothesis model based on the Knowledge Construction Model and refine it through feedback from experienced researchers to ensure validity and reliability.

B. **Competence Story**—*Maximizing Skills and Potential* (Intelligence): While extensive literature exists on research evaluation methods, the role of research activities in fostering knowledge synthesis remains underexplored. This study will evaluate the significance of these activities by analyzing survey data using covariance structure analysis. Conducting this analysis requires proficiency in statistical software and an understanding of advanced research methodologies.

C. **Connection Story**—*Considering Social and Environmental Factors* (Involvement): To collect comprehensive data, I will survey management and materials science researchers, asking them to rate the importance of various research activities on a scale of 1 to 5. To ensure broad participation, I will collaborate with senior researchers to distribute the questionnaire and consult young researchers to validate the relevance of the survey questions.

III. **Stage 3—Integration:**

1. **Feasibility Story**—*A Culturally Feasible Plan* (Integration of Intention and Connection Stories): Many researchers are interested in identifying key research practices that facilitate knowledge synthesis and enhance research outcomes. To gather empirical data, I will survey management and materials science researchers and ask them to evaluate the importance of various research activities. To ensure broad participation, I will collaborate with senior researchers to distribute the survey and consult young researchers to refine the clarity and relevance of the research questions.

Explanation: This story integrates key aspects from both:

- **Intention Story**: Focuses on understanding critical research practices, promoting knowledge synthesis, and ensuring research validity through expert feedback.
- **Connection Story**: Emphasizes practical data collection methods, including surveying specific researcher groups, leveraging senior researchers for distribution, and engaging young researchers for validation.

The Feasibility Story merges the conceptual focus of the Intention Story with the methodological framework of the Connection Story, creating a cohesive and actionable research plan.

2. **Desirability Story**—*A Systemically Desirable Plan* (Integration of Competence and Connection Stories): Research examining the impact of research activities on knowledge synthesis remains a relatively unexplored area. This study addresses this gap by investigating how specific research activities contribute to knowledge

synthesis. To achieve this, I will use covariance structure analysis on survey data from management and materials science researchers.

Explanation—This story synthesizes insights from both:

- **Competence Story**: Highlights the novel focus on research activities in knowledge synthesis, addressing an academic gap, and applying covariance structure analysis with established statistical tools.
- **Connection Story**: Incorporates the data collection approach, specifically surveying management and materials science researchers.

The Desirability Story combines the Competence Story's theoretical foundation and analytical approach with the Connection Story's practical survey methodology, forming a cohesive research framework.

3. **Strategy Story**—*A Synthesized Plan* (Integration of Feasibility and Desirability Stories): This research aims to identify specific research activities and combinations that foster knowledge synthesis. To accomplish this, I will:

   - Develop a questionnaire based on the Knowledge Construction Model.
   - Collaborate with external researchers to distribute and validate the survey.
   - Analyze collected data using covariance structure analysis to determine key factors contributing to knowledge synthesis.
   - Compare findings across disciplines (management and materials science) to examine disciplinary differences in research practices.

   This research will generate significant interest among academics and students, providing valuable insights into effective research methodologies.

Explanation—This final story integrates and extends elements from both:

- **Feasibility Story**: Contributes the focus on knowledge synthesis, the survey methodology, and collaboration with external researchers for validation and distribution.
- **Desirability Story**: Emphasizes the role of research activities in knowledge synthesis, using covariance structure analysis and targeting management and materials science disciplines.

The Strategy Story unifies these aspects into a comprehensive research strategy and expands upon them by introducing:

- A structured questionnaire development process based on the Knowledge Construction Model.
- Comparative analysis across disciplines to examine differences in research activities.
- A focus on the broader academic significance of the study.

Thus, the Strategy Story serves as the final synthesis, building upon and extending the core ideas from the Feasibility and Desirability Stories, resulting in a well-rounded research narrative.

Correction to: Decision-Making with the Intuitive Mind

Correction to:
Y. Nakamori, *Sensible Decision-Making*,
https://doi.org/10.1007/978-981-96-6811-3_2

The author provided corrections to the equations in chapter 2 of the book, which had been inadvertently published with incorrect equations. The initial version has now been corrected.

The updated version of this chapter can be found at
https://doi.org/10.1007/978-981-96-6811-3_2

Y. Nakamori, *Sensible Decision-Making*, https://doi.org/10.1007/978-981-96-6811-3_7

Index

Y. Nakamori, *Sensible Decision-Making*, https://doi.org/10.1007/978-981-96-6811-3

MIX
Papier aus verantwortungsvollen Quellen
Paper from responsible sources
FSC® C105338

If you have any concerns about our products,
you can contact us on
ProductSafety@springernature.com

In case Publisher is established outside the EU,
the EU authorized representative is:
Springer Nature Customer Service Center GmbH
Europaplatz 3, 69115 Heidelberg, Germany

Printed by Libri Plureos GmbH
in Hamburg, Germany